S. I. ROSEN

*University of Maryland*

# INTRODUCTION

# TO

# THE

# PRIMATES

*living and fossil*

PRENTICE-HALL, INC., ENGLEWOOD CLIFFS, NEW JERSEY

*Library of Congress Cataloging in Publication Data*

Rosen, Stephen I.
    Introduction to the primates.

    Bibliography:  p.
    1. Primates.  2. Primates, Fossil.
3. Human evolution.  I. Title.
QL737.P9R67          599'.8          73–14802
ISBN   0–13–493460–1

*© 1974 by Prentice-Hall, Inc., Englewood Cliffs, New Jersey*

PRINTED IN THE UNITED STATES OF AMERICA

10 9 8 7 6 5 4 3 2 1

PRENTICE-HALL INTERNATIONAL, INC., *London*
PRENTICE-HALL OF AUSTRALIA, PTY. LTD., *Sydney*
PRENTICE-HALL OF CANADA, LTD., *Toronto*
PRENTICE-HALL OF INDIA PRIVATE LIMITED, *New Delhi*
PRENTICE-HALL OF JAPAN, INC., *Tokyo*

In memory of David

# Contents

# *Preface*

This book is written particularly for the beginning student who is being exposed for the first time to the *Primates* as a field of academic study. This student may be in an anthropology, primatology, zoology, psychology, or natural science course, since the study of the primates knows no boundaries. The genetics student may also find this book of interest in light of the great variety of primates, living and fossil. Whatever the case may be, the student is faced with a mass of different types of primates and an even greater mass of what must appear miscellaneous information. This introductory text is designed to simplify this situation, for it should be the goal of every beginning text to simplify the complex.

First, the *Primates* are divided into two logical groupings: the modern, or living primates; and the fossil ones, the primates of the past. A second way of simplifying matters is the use of the **pattern concept**— no single primate or group of primates is defined on a basis of one or two characteristics but on the general pattern displayed. By use of boldface type, the student is presented not only with what is important and what to remember but also has a quick reference system for reviewing. Each chapter has a pattern summary at the end so the reader can check his recall and quickly refer back to the corresponding boldface print.

Complex subject matter need not be presented in a complex manner; students in my classes who successfully used the above method urged me to write this book in order to share this approach with fellow students elsewhere. The study of the Primates is a most fascinating and rewarding adventure. This interdisciplinary science is worthy of the attention of our best students.

S. I. ROSEN

# PART I

# THE  LIVING  PRIMATES

# Chapter 1

# We the Primates

No one would deny that modern man, *Homo sapiens sapiens,* is a primate. Yet few people understand why man is classified with these "animals" or he is related to them. Even anthropologists, who study him more than any other group of scholars, cannot adequately define man. Neither man nor his fellow primates, living and fossil, can be defined by a few features. Rather, they are defined by complexes of characteristics—that is, **patterns**. The great British anatomist and physical anthropologist, W. E. LeGros Clark, pioneered this approach in what he called the **total morphological pattern.**

Primates are **metazoans**—that is, multicellular animals. Because they possess an internal skeleton, they are also called **chordates**. They also have a segmented vertebral column (spine) instead of a straight rod and thus are classified as **vertebrates**. We, the primates, are also called **homeotherms**, since like our fellow mammals and birds, we are able to maintain constant body temperatures within a few degrees in most climatic conditions. We are **mammals** because we possess a complex of traits such as: mammary glands (breasts), suckling of our young, a hairy body covering, giving birth to living young rather than laying eggs, and so on. Thus, we the primates are metazoans, chordates, vertebrates, homeotherms, and mammals. Yet so are hundreds of other animals, and they are not primates.

What makes us primates? Part of the answer lies within an under-lying principle—as an order of animals, we the primates have retained a rather **primitive and generalized anatomy**, that lacks a great number of specializations except perhaps neurological development. Our anatomy is "primitive" in the sense that it has not radically changed from the earliest mammals, especially those ancestral to the primates. Dr. Adolph H. Schultz, the foremost student of the primates, has stated how remarkable it is that beneath the skin most primates are quite uniform and conservative in form compared with other animals. In the primates there is a tremendous number of variations on the same theme, or **primate pattern**.

### *The Primate Pattern*

Primates have an anatomy that enables them to maintain **erect** and **semierect postures** and **locomotor patterns**. The primates maintain such postures more habitually than most other animals, except for birds. Through "natural selection," nature has required the primates to have varying abilities to sit and stand erect on the ground and in trees, thus freeing the hands for investigation and manipulation of the environment. This freedom has allowed the hands to bring objects to the nose for smelling and to the mouth for tasting, lessening the need for a large smelling and food-gathering apparatus (a muzzle) for most of the pri-mates. The anatomist F. Wood Jones called this freedom the **emancipa-tion of the forelimbs**. In pronograde animals (those who walk on "all fours") the forelimbs are used as structures of propulsion and support. All primates are capable of standing erect (orthograde posture), and many are capable of bipedal movements such as running, springing, leaping, hopping, and walking (Fig. 1–1).

A very primitive trait that survives in our order is **pentadactylism** —having five digits on each hand and foot rather than a webbed struc-ture such as a paw. The horse once had five digits but now shows only remnants of them in the internal structure of its hoof. Pentadactylism allows primates a good grip for grasping tree trunks and branches and for manipulating food and other objects in its environment. In fact, primate hands have changed very little in the evolutionary history of the order. The primates retain a very archaic functional axis in their hands. The third finger is the primate functional axis, except in the lemur and loris, which have the axis running through the fourth finger.

*Fig. 1–1* A female pygmy chimpanzee (*Pan paniscus*) in a semierect posture Note the grasping right foot and the sexual skin area extending from the buttocks. (San Diego Zoo Photo by Ron Garrison.)

Instead of the typical mammal claws, **flattened nails** are found on the ends of primate digits. The tree shrews are an exception in having only claws. As one approaches the higher primates, the nails become more human in character. The flattened nail permits a grasp suited for moving or swinging from branches and also allows the primate considerable tactile sensation (Fig. 1–2).

As we go up the primate ladder toward man, we find that the relative **density of hair** on the primate body becomes progressively **less**. It has been found that the individual hair diameters become progressively larger also. Thus we find thicker hairs but fewer of them per square inch of body surface. All this being true, man is still far from being a "naked ape." Higher up the ladder, **fewer special** tactile **hairs** (vibrissae—similar to cat and dog whiskers) are found; the apes and man have none.

*Fig. 1–2* Grasping primate hands. Note the increasing opposability of the thumb to index finger, as one approaches man. (After Pfeiffer, John E., *The Emergence of Man.* Copyright 1969, Harper and Row.)

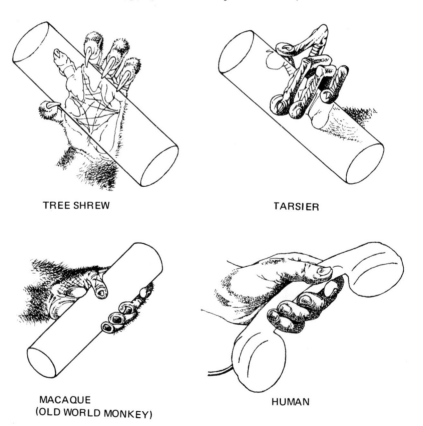

TREE SHREW

TARSIER

MACAQUE
(OLD WORLD MONKEY)

HUMAN

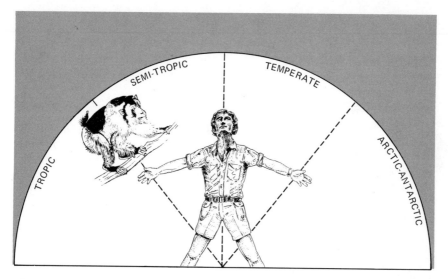

*Fig. 1–3* While the nonhuman primates are largely limited to tropical climates, man through his cultural and biological plasticity has been able to adapt to a multibiome. (S. I. Rosen.)

Primates usually inhabit **tropical and semitropical climates**. There are exceptions, however, such as the Japanese snow macaques and man, who through his culture acclimates himself to all types of geography (Fig. 1–3).

A **great variety** of **body sizes** and **weights** is displayed in the primate pattern. A good zoo might exhibit a mouse lemur weighing several ounces within a few yards of a 300-pound gorilla. As we go up the primate scale, the species are generally larger and heavier.

Compared with other mammals, we the primates cannot smell things very well. There has been an evolutionary **reduction** of the **olfactory**, or "smell" **area** of the **brain**. As a result of this reduction, there has been a **reduction** of the **snout** or muzzle, so that most of the higher primates have a face and a simple dry nose. Most often where evolution has changed the structure of the brain, the gross anatomic part which that area of the brain affects will also be changed in its structure. While the olfactory area was being reduced, the **visual area** underwent **expansion** (Fig. 1–4). These two major changes are likely results of the **adaptation** of most primates to an **arboreal** (tree-dwelling) **environment**. It is more important for a South American spider monkey to be able to see the next tree limb he is swinging to than to smell it. To fully function in an arboreal niche and still be a primate, one must also have **stereoscopic vision**. With reduction of the snout, the eyes became

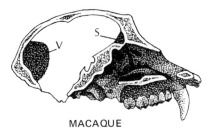

TREE SHREW

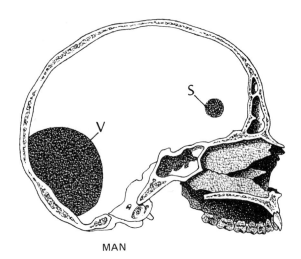

MACAQUE

*Fig. 1–4* The primate trend toward reduction of the smell (*S*) area and the increase of the vision (*V*) area of the brian, as revealed diagrammatically in sagittal (half) sections of three primate skulls. (After Pfeiffer, John E., *The Emergence of Man,* Copyright 1969, Harper and Row.)

MAN

V-VISION
S-SMELL

positioned closer together, creating visual overlap in order to perceive depth. With most primates being active in the daytime—that is **diurnal**— a further result has been the acquisition of **color vision.** Clearly primates are "eye creatures."

Some mammals are able to pick up objects and hold them between their hands or paws; this is called convergent prehensibility. Most mammals use their elongated muzzle and dental apparatus with some aid from claws in such activities. The primates are unique in that they possess varying abilities to hold things between their thumbs and one or more digits and likewise their big toe. Thus pseudo and true **opposability** of the **thumb** and **fingers** is a hallmark of the primates. As part of the adaptation to an arboreal environment, primates developed grasping hands and feet for climbing; these structures have changed very little in most of the primates. Dr. John Napier has classified the prehensile grips of primates into two categories: the **power grip** and the **precision grip** (Fig. 1–5). A key to these abilities is the fact that the thumb (pollex) and the big toe (hallux) to varying degrees function independently of the other digits and are widely separated from them.

*Fig. 1–5* The two basic primate grasps: (a) the power grip and (b) precision grip. (J. Hourihan.)

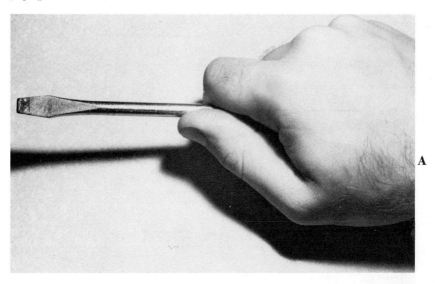

A

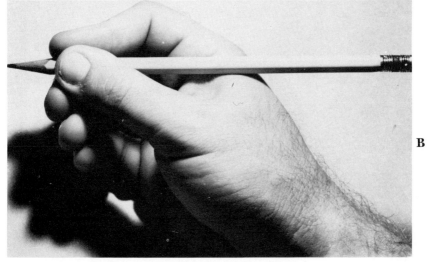

B

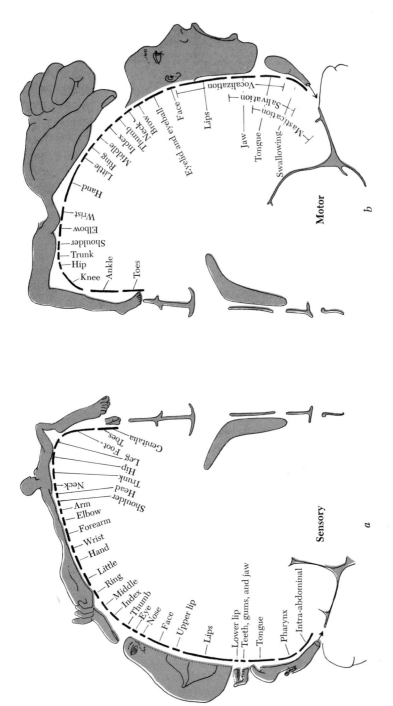

*Fig. 1–6* A diagram of a section through a human brain. The primate cerebral cortex has unusually large amounts of surface area devoted to the hand, especially the thumb, and also to the vocal organs. (After Penfield and Rasmussen, *The Cerebral Cortex*, The Macmillan Co., 1950.)

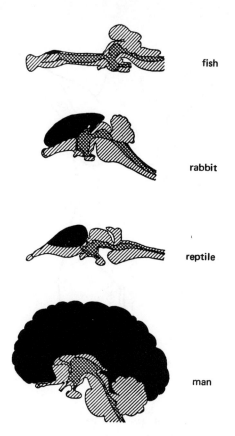

fish

rabbit

reptile

man

*Fig. 1–7* Brain sagittal sections of representative vertebrates showing the differences in size and complexity of the cerebral cortex (black). (After Magoun, in *Evolution After Darwin*, ed. S. Tax, University of Chicago Press, 1960.)

In fact, most primate digits show independent movement. Neurological changes have also taken place; the association and motor areas for the thumb take up a great portion of the primate brain's cerebral cortex relative to other motor-skill areas (Fig. 1–6).

As a group, we the primates are considered to be the most intelligent of all animals. For body size and weight, primates possess **relatively large and complicated brains**. These brains are more complex in their gross and microscopic anatomies than those of any other order. The dramatic showcase for this complexity is the convoluted (folded) cerebral cortex, which integrates sensory information and selects motor responses to such information (Fig. 1–7).

The male mammal has a penis which is usually supported by a slender bone, sometimes called a baculum, or *os penis*. This bone is actually a muscle area that has turned to bone. Except for man, most male members of our order possess this bone. In addition to the baculum, a flap of skin usually attaches the mammal penis to the abdomen. In the male primates, this skin flap is absent or so reduced that they have a **pendulous penis**, a trait they share only with bears, bats, and hyenas.

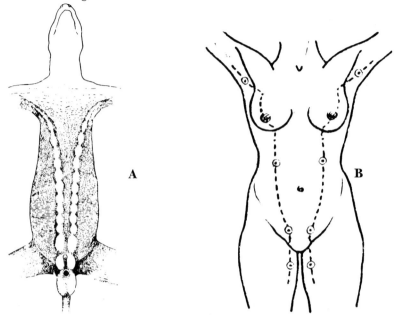

*Fig. 1–8* The primitive mammalian milk lines in a hypothetical mammal (A) and the typical primate pattern of two pectoral mammary glands (B). (A—after Wood Jones)

Anyone who has seen a mother dog or pig knows that female mammals (males also) usually have multiple mammary glands. These breasts can occur anywhere on the ventral (underside) surface of the body along two areas called the mammalian milk lines (Fig. 1–8). Most primates, along with sea cows and elephants, have only **two breasts**. They are usually located in a **pectoral** (chest) position.

Except for a few cases, primates do not have litters; **single births** are the general rule. Most primate females possess a simplified uterus, that is called a **unicornate** (one-horned) **uterus** while other mamals have bicornate or double-horned uteri, which are a factor in multiple births. The primates also possess a unique placenta, or fetal membrane, which we unfortunately do not understand very well. A reduced number of offspring at any one time permits the female great freedom of movement in both terrestrial and arboreal environments. It also assures a greater chance of survival for the infant, whereas multiple siblings have a greater chance of perishing. Primates lack the home-base nesting locations that many mammals with litters maintain.

Generally the primates do not exhibit seasonal sex, although some species do tend to have more births at certain times of the year. The latter may be related to climate, food supply, and ecological factors, since most primate females ovulate twelve times during the year. This **year-round fertility** is probably one key to the great biological success of the primates as an order. The continuous introduction of new infants into the group helps to stabilize population size.

Primate infants have **prolonged physical and emotional dependence** upon their mothers; coupled with this are prolonged growth and maturation periods and long life spans compared to other animal groups (Fig. 1–9). These prolonged periods insure the safety of the single infant

*Fig. 1–9* The trend toward prolonged growth and maturation periods in primates. (Courtesy of A. H. Schultz.)

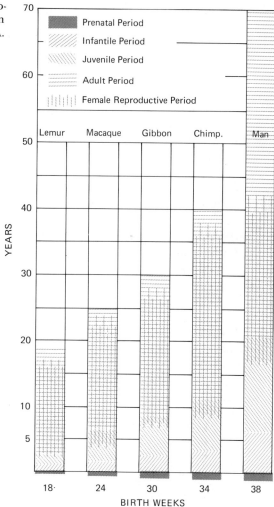

and also allow time for complex learning and the establishment of complex social relationships.

In the case of the teeth, primates have four upper and four lower incisor (front) teeth; these teeth are relatively larger than those of most mammals. The molar (cheek) teeth have quite simple grinding surfaces, unlike the very complicated surfaces found on the molars of vegetarian mammals such as the horse. Generally the primates show a **reduction in the number of teeth** and **lack of dental specialization**, thus we the primates have a generalized dentition.

For a long time most of the primates, except man, have been considered to be vegetarians. In recent years primatologists have discovered that some primates such as the baboon and chimpanzee occasionally capture, kill, and eat small game in the wild and that large numbers of other primates obtain animal protein by eating lizards, insects, small birds, and eggs. In our zoos and primate centers, we have found that primates can acquire a preference for meat after being introduced to the diet. Although quite a few primates have evolved very restricted dietary preferences, as a group they are best described as being **omnivorous.**

All primates exhibit traits due to arboreal adaptation. In fact, a great many of the pattern traits we have discussed are such adaptations. Some others of importance are: **retention of the primitive clavicle** (collar bone), which most mammals do not possess and which acts as a strut or brace for the shoulder girdle; a separate radius and ulna (forearm bones); and a separate tibia and fibula in the leg (except in the case of the tarsier). A wide range of shoulder movements, a reduction in the length of the external tail to the point where it is absent in the apes and man, and a lengthening of the limbs together with a shortening of the vertebral column are also primate skeletal trends.

None of the above traits characterizes all members of the order *Primates*. There are always exceptions to single parts of the pattern; that is why one must think of patterns or complexes rather than single traits.

Even with the accumulated knowledge of modern primatology, we find the classic definition of our order remains the best one to date:

> Unguiculate, claviculate placental mammals, with orbits encircled by bone; three kinds of teeth, at least at one time of life; brain always with a posterior lobe and calcarine fissure; the innermost digits of at least one pair of extremities opposable; hallux with a flat nail or none; a well-developed caecum, penis pendulous; testes scrotal; always two pectoral mammae. (St. George Mivart, 1873, Zoological Society of London.)

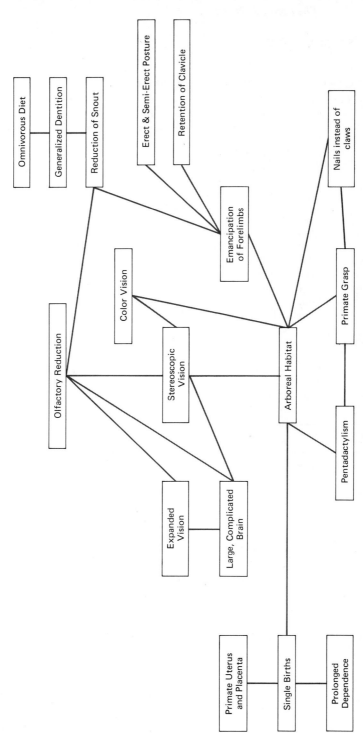

*Fig. 1–10* The more obvious interrelationships in the primate pattern. (S. I. Rosen.)

# SUMMARY

# Chapter 2

# Classification of Primates

## taxonomy in turmoil

Knowingly or not, we are all taxonomists—classifiers of things, living and inanimate. We put the elements of our lives into convenient categories, the cubbyholes of the mind. Biological scientists do the same thing but call it **systematics** or **taxonomy**,[1] the science of classification. The first recognized taxonomist (he was the first in many fields) was Aristotle (4th century B.C.). Aristotle's classification system was hierarchical; he graded animal forms into ideal degrees of superiority and complexity. Naturally he placed man at the top. Much of the early classification of animals was based on the Platonic idea of the **archetype**, an ideal form of organization. Animals were classified as to how closely they met the archetype, even though this ideal need not have really existed in nature.

Taxonomy had to wait until the eighteenth century to become a true science. The man who brought the discipline to maturity was a Swedish botanist and naturalist named **Carolus Linnaeus** (1707–1778). In the tenth edition of his book, the *Systema Naturae*, Linnaeus classified

[1]Modern biology defines these words differently. Systematics is considered the study of how animals are similar and dissimilar, while taxonomy is the study of theories of classification. We will consider them to be synonymous.

the primates. In fact he named our order *Primates,* meaning "the first." His classification formed the groundwork for modern primate taxonomy. Realizing that the main purpose of classification should be communication, Linnaeus devised a taxonomic language termed **binomial nomenclature**, meaning "two-word system." In this system each animal has a two-word (Latinized) name by which scientists, whatever their native language, can recognize that animal. The two categories are **genus** and **species**. For example, modern man is a member of the genus *Homo* and the species *sapiens*.

Linnaeus also created three other taxonomic categories, or levels: kingdom, class, and order. He named our order *Primates* and included in it: man, apes, monkeys, lemurs, bats, and "flying lemurs." The last two animals are no longer classified as primates; the latter is not even a lemur. Linnaeus's classification stands to this day as the basis for modern primate taxonomy:

| Category | Taxon |
|---|---|
| Kingdom | *Animalia* |
| (Subkingdom) | *(Metazoan)* |
| Phylum | *Chordata* |
| (Subphylum) | *(Vertebrata)* |
| Class | *Mammalia* |
| Order | *Primates* |
| Genus | *Homo* |
| Species | *sapiens* |

The Linnaean system was based on anatomical features alone. Most classification is still morphological (anatomical) in nature although there are trends towards using the behavior, biochemistry, physiology, and molecular genetics of an animal as bases for classification into a specific taxon (category).

Linneaus's classification of the primates was not at first completely accepted by the scientific community of his time, especially because it employed as a basic idea the close anatomical relationship between man and the anthropoid apes. A German anatomist, sometimes called the "father of physical anthropology," Johann Friedrich Blumenbach (1752–1840), in 1791 classified the primates into two orders: the *Bimanus* (two-handed) and the *Quadrumana* (four-handed). Man was placed in the *Bimanus* and the apes, monkeys, and lemurs in the *Quadrumana*. These two orders are not recognized today, although they did have considerable meaning for the eighteenth-century scientific and intellectual communities.

*Important Concepts*

Unlike their predecessors, modern taxonomists classify **populations** of animals, not single animals. These populations are natural breeding units that are defined by anatomical and geographical boundries. Populations are also called on a large scale **species** and on a smaller scale **subspecies** or races. A group of related species composes a **genus**. When considering the living primates, the most important and meaningful taxonomic level is the species; in fossils it is the genus. This appears simple and straightforward, but there is a major problem—the definition of a species. All biologists do not adhere to the same definition. The most common definition of a species is:

> a population of animals which interbreed and produce viable (living) and fertile offspring.

Some biologists claim that the animals do not necessarily have to interbreed but must be anatomically capable of doing so; thus they remain potential interbreeders. Some stipulate that to be potential interbreeders the populations must be geographically close enough to breed. Other taxonomists hold that the populations must actually and naturally interbreed to be the same species; by this standard, laboratory breedings of animals whose natural geographies are not close are not valid taxonomic experiments. Even with this confusion, the species level is the most well defined of all taxonomic categories.

Within each species, there is usually a moderate amount of variation in anatomical features, especially in the more superficial ones such as hair color and pattern. These variations tend to correspond roughly to differences in geographic distribution and are recognized by putting different members of the same species into subspecies, or races. These variations often show rather orderly gradations that correspond approximately to the geographic distribution of the populations. We call the gradation for a specific trait a **cline**. For example, there may be a cline for the amount of yellow hair in the coats of baboons from one part of Africa to another. Classification of primates into subspecies is thus largely based upon minor variations and geographical distributions. Because such variety can exist within a species, few species are considered to be "average." The presence of anatomical variation and the possibility of geographic (and therefore reproductive) isolation makes each subspecies an **incipient** or potential, **species**.

The other levels of classification lack clear definition; assignments to superfamilies, families, subfamilies, and so on is very arbitrary.

Linnaeus had only five taxonomic levels; today there are more than twenty. Having so many levels has led to an unusual amount of confusion in classifying the order *Primates*. There are approximately 200 species and 55 genera (pl. of genus) of primates that are recognized by various authorities.

## *"Lumpers" and "Splitters"*

Taxonomists often refer to their fellow workers as either "lumpers" or "splitters." The lumpers tend to classify many animals into few taxa (pl. of taxon), while splitters prefer to divide animals into many separate taxa. While most taxonomists are neither exclusively one type or the other, the modern trend is to "lump" in order to simplify matters and to compensate for past exaggerations of minor variations in single traits. The same tendency is found in primate taxonomy due to the fact that most of our primate classifications come from the discoveries and observations of hunters, animal collectors, and early naturalists of the seventeenth through nineteenth centuries. These men had little knowledge of the diversity of primate life. Sadly, most of their species designations were based on fur colors and patterns. Almost every discovered (and rediscovered) primate was given a new classification.

The scientific naming of animals is based partly upon what is called the **law of priority**. This rule holds that the first classification (Latin name) given to a type of animal is its only classification. Thus, a new name cannot be given to an animal that has been described previously. Unfortunately, this rule has been broken often by students of the living and fossil primates. Too many primate subspecies have been made "new" species and species made genera.

## *The Evolutionary Aspect*

Linnaeus and other eighteenth-century taxonomists did their work before the idea and theory of evolution became widely known, which was in the second half of the nineteenth century. Their classifications were generally hierarchical but not evolutionary in nature. Many modern taxonomists believe that the classification of an order of animals should reflect the order's **phylogeny**—the evolutionary history of the group. Other taxonomists hold that taxonomies should reflect only anatomical similarities and dissimilarities; if the classifications happen to also reflect phylogenies, this is after the fact. The primates present no problems

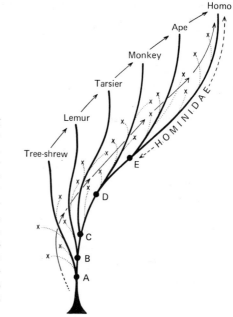

*Fig. 2–1*  Diagram showing that the living primates form a series from treeshrew to man, a series that suggests a general trend of evolutionary development. From comparative studies of such types it is possible to postulate the probable linear sequence of human ancestry. This sequence leads from the basal primate stock through hypothetical transitional stages, A–E, at which the different ramifications of the primates are presumed to have branched off. Fossil primates approximating to the postulated ancestral stages are presented by crosses. (After LeGros Clark, 1959.)

here since the classifications to a great degree reflect actual and theoretical aspects of the evolutionary history of the order (Fig. 2–1). T. H. Huxley, patriarch of the famous and brilliant Huxley family of England and defender of Charles Darwin, stated it so well:

> Perhaps no order of mammals presents us with so extraordinary a series of gradations as this—leading us from the crown and summit of the animal creation down to creatures, from which there is but a step, as it seems, to the lowest, smallest, and least intelligent of the placental *Mammalia* (1863).

The study of an order with so much variability as the *Primates* can present considerable problems. If you find two primate populations which appear very similar, are they closely related species? They need not be. They may simply be the products of **parallel evolution**—two types of animals which had a distant common ancestor many millions of years ago but went separate evolutionary ways, yet evolving parallel adaptations. Due to a common genetic inheritance and adaptation to similar environments, they appear superficially similar today. The New World monkeys and the Old World monkeys are a likely example of parallel evolution. Such cases can occasionally also be explained by what we call **convergent evolution**. In this case, animals that do *not* have a common ancestor, but are in very similar if not the same environments, evolve similar adaptations. A good example of convergent evolution is some of the primitive arboreal marsupials (a protomammal group) and the prosimians (lowest of the primates). Superficially some of the members of these two groups look much alike; they have made similar arboreal adaptations to rather narrow econiches. They have "converged" on similar environments. The concepts of parallel and convergent evolution are often overused to explain certain living and fossil nonhuman primates but are rarely applied to the fossil record of man.

It is clearly evident that all animals are products of the evolutionary history of their order and of their present environment. A key to understanding both living and fossil animals is the concept of **natural selection**, the major process, or means, of evolution. The natural environment changes at different times and at different rates; survival in the changed environment may require new anatomical traits. Nature blindly "selects" for certain traits. A trait that enables an animal to survive and thus reproduce its own kind is an **adaptive trait**. Occasionally an animal will possess an anatomical trait that will be advantageous in a future environmental setting. Such traits that yield future survival benefits are termed **preadaptive traits**. Some traits may not yield survival benefits at all and might even be disastrous for the animal; these are **nonadaptive traits.**

The phrase "survival of the fittest" is one most every student has heard and even used, yet is too often misunderstood. Many people believe this idea means that the strong survive, and thus the strong are the fittest. This is often furthest from the actual truth. Modern biology speaks instead of **Darwinian fitness**, which is measured by the ability of a species to not only survive but to reproduce enough of its own species to insure the continuity of the species through time. Therefore, Darwinian fitness is actually "reproductive fitness." Obviously an animal must survive at least till early reproductive maturity in order to reproduce. Since nature can select for different traits at various times by means of environmental changes, it stands to reason that the more variety a group of animals possesses, the greater its chances of meeting the test of change. The *Primates* are a very old and quite successful order of mammals (see Fig. 9–1, Chapter Nine). A prime key to their success is that the order contains a large number of species that exhibit significant amounts of subspecific variation; we call these **polytypic species.** The more variety, the greater the chances of having what is needed for survival. These numerous types, or varieties, have yielded great survival benefits for the order; yet with all this variety, most primates have rather primitive and generalized (nonspecialized) anatomies. A generalized anatomy can adapt to many different situations. The generalized primate hand is an excellent example of this. This type of hand can be used for grasping objects and food, for locomotion on the ground and in the trees, and also for protection. The horse's hoof, a very specialized structure, is primarily designed for terrestrial locomotion, secondarily for defense. A generalized, unspecialized anatomy is usually found early in the history of a mammalian order, and thus is a primitive, or more properly, archaic anatomy. The primates have kept very close to the original primate ¬natomical pattern. Even the highly complex and seemingly specialized primate brain is generalized in that is allows for a great variety of noninstinctual responses to situations. The neurological systems of other mammals are specialized for a certain number of precise responses. The unspecialized mammal will always have the greater advantage unless the environment becomes uniformly narrow. In the history of the vertebrates, overspecialization has almost always spelled extinction.

All of these concepts are fundamental to understanding **evolution,** which is simply change through time. It is an ongoing process; evolution does not stop, although it may slow down to the point where it is nonapparent. Evolution also has the property of **irreversibility**: once an anatomical trait is lost, it is not usually regained, at least not in identical form.

### The Turmoil

Taxonomy was originally designed to make it easy for all students, old and new, to identify animals. In the case of the primates, it has become a confusing game of oversplitting and occasional mislumping. The nomenclature is especially confusing in the case of the fossil primates; here almost every rule has been broken too often and too regularly. It is such a bewildering mass of names and synonyms that the student (and sometimes the scholar) is easily lost. The primate taxonomy being in such disorder, some primatologists hold that the most scientific designations for some primates are their common names.

While primate taxonomy is supposedly based on the recognition of total morphological patterns, very frequently only one or a few traits are taken into account in classification. Many primate species and some genera are based on fur color and pattern. Too many hybrids have been found in nature and the laboratory.

Because of the many problems inherent in primate classification, most zoologists and taxonomists steer clear of this hotbed, particularly the fossil primates. This has left primate taxonomy in the hands of a few physical anthropologists and anatomists who often are not trained in taxonomy. As more biologists become involved in the study of primates, the taxonomy of this order will likely become more logical and usable, especially against a background knowledge of nonprimate animals that have been more adequately studied. Unfortunately, until that day arrives, the student new to primate biology carries the unfair burden of learning his teacher's mistakes.

## Major Primate Groupings

### PROSIMII (HALBAFFEN) AND ANTHROPOIDEA (SIMIAE)

The order Primates is generally divided on a large scale into two logical groups or suborders, the *Prosimii* and the *Anthropoidea*, a classification Dr. G. G. Simpson is largely responsible for. The **prosimians** are an ancient and rather nontypical group of primates; as a group, they exhibit features that deviate from the primate pattern more often than any other primate group. Paleonotologically, they are very old (at least seventy million years) and are closer to the nonprimate-primate boundry than any other primate group. European primatologists generally refer to the prosimians as the **Halbaffen** or "half-monkeys." The prosimians are also often termed the **lower primates.**

The **higher primates** are the *Anthropoidea*; this group is composed of **monkeys, apes,** and **man.** The word "anthropoid" usually means "ape" to most people, yet man and monkeys are as much anthropoids as the chimpanzee and the gorilla. Again, European scholars use a different word for anthropoid, in this case *Simiae*, from which comes the word "simian." Orginally the *Simiae* were Old World macaque monkeys and the orang-utan, a great ape.

### PLATYRRHINES AND CATARRHINES

Some students of the primates divide the Anthropoidea into two groups, the **Platyrrhines** and the **Catarrhines.** Both of these terms refer to the external shape of the noses of New World and Old World anthropoids. The platyrrhines ("flat-nosed"), or **New World Monkeys** have a broad, fleshy nasal septum and round nostrils, while the catarrhines ("downward-nosed"), Old World primates (monkeys, apes, and man, but not prosimians), have narrow, fleshy nasal septa (pl. of

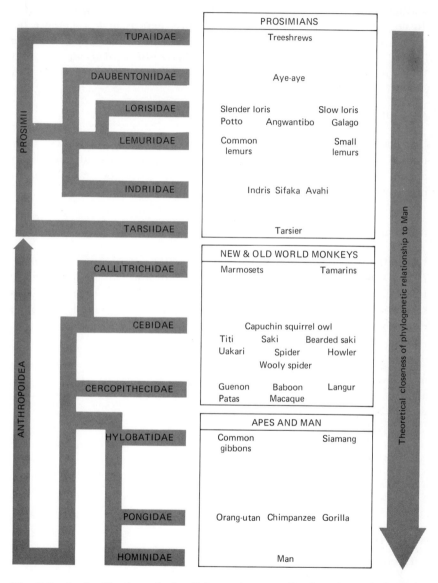

*Fig. 2–2* A classification of the living primates reflecting theoretical phylo-genetic closeness of different types. (Modified after Napier, J. R., "Prospects in Primate Biology," *Proc. U.S. National Museum*, 1968.)

septum) and comma-shaped nostrils (see Fig. 5–1). Many primatologists do not like these terms because they imply there is little difference between these two major groupings except for the nose. The taxon **Ceboidea** is used for the New World monkeys, **Cercopithecoidea** for the Old World monkeys, and **Hominoidea** for the apes and man.

ANTHROPOID APES

The apes are generally placed in two distinct families. One family is the **Hylobatidae** or gibbons, occasionally called the "lesser apes." The other family is the **Pongidae**, or great apes, which consist of the orang-utan, chimpanzee, and gorilla.

HOMINIDS

Egocentrically, men have sought to place the living species of man, *Homo sapiens sapiens,* and his ancestors into a special family separate from the great apes. This family is the **Hominidae**, or hominids. This separation from the great apes can be challenged on grounds that man does not represent a unique primate but is a logical variation of the primate evolutionary pattern.

A CLASSIFICATION OF LIVING PRIMATES

**SUBORDER TUPAIOIDEA**

FAMILY TUPAIIDAE

| Genus *Tupaia* | common, or large, treeshrew | Southeast Asia |
| Genus *Dendrogale* | smooth-tailed treeshrew | Southeast Asia |
| Genus *Urogale* | Philippine treeshrew | Philippines |
| Subfamily Ptilocercinae | | |
| Genus *Ptilocerus* | pen-tailed (feather-tailed) treeshrew | Southeast Asia |

**SUBORDER PROSIMII**

**Infraorder Lemuriformes** Madagascar

FAMILY LEMURIDAE

| Genus *Lemur* | lemur (true lemur) |
| Genus *Hapalemur* | gentle lemur |
| Genus *Lepilemur* | sportive (weasel) lemur |
| Genus *Cheirogaleus* | dwarf lemur |
| Genus *Microcebus* | mouse lemur |

FAMILY INDRIIDAE Madagascar

| Genus *Indri* | indris |
| Genus *Propithecus* | sifaka |
| Genus *Avahi* | avahi |

FAMILY DAUBENTONIIDAE      Madagascar
     Genus *Daubentonia*      aye-aye

**Infraorder Lorisiformes**
     FAMILY LORISIDAE

| | | |
|---|---|---|
| Genus *Loris* | slender loris | South India, Ceylon |
| Genus *Nycticebus* | slow loris | Southeast Asia |
| Genus *Perodicticus* | potto | Africa |
| Genus *Arctocebus* | angwantibo (golden potto) | West Africa |

     FAMILY GALAGIDAE

| | | |
|---|---|---|
| Genus *Galago* | galgo, or bushbaby (includes dwarf species) | Africa |

**Infraorder Tarsiiformes**
     FAMILY TARSIIDAE

| | | |
|---|---|---|
| Genus *Tarsius* | tarsier | Southeast Asia |

## SUBORDER ANTHROPOIDEA

**Infraorder Platyrrhina**
Superfamily Ceboidea
     FAMILY CALLITHRICIDAE

| | | |
|---|---|---|
| Genus *Callithrix* | marmoset | South America |
| Genus *Cebuella* | pygmy marmoset | South America |
| Genus *Saguinus* | tamarin | Central and South America |
| Genus *Leontideus* | golden lion tamarin | South America |

     FAMILY CALLIMICONIDAE

| | | |
|---|---|---|
| Genus *Callimico* | Goeldi's marmoset | South America |

     FAMILY CEBIDAE
     *Subfamily Aotinae*

| | | |
|---|---|---|
| Genus *Aotes* | owl (night) monkey, Douroucouli | Central and South America |

     *Subfamily Pitcheciinae*

| | | |
|---|---|---|
| Genus *Pithecia* | saki | South America |
| Genus *Chiropotes* | bearded saki | South America |
| Genus *Cacajao* | uakari | South America |

     *Subfamily Cebinae*

| | | |
|---|---|---|
| Genus *Cebus* | capuchin monkey | Central and South America |
| Genus *Saimiri* | squirrel monkey | Central and South America |

     *Subfamily Alouattinae*

| | | |
|---|---|---|
| Genus *Alouatta* | howler monkey | Central and South America |

     *Subfamily Atelinae*

| | | |
|---|---|---|
| Genus *Ateles* | spider monkey | Mexico, Central and South America |
| Genus *Brachyteles* | woolly spider monkey | South America |
| Genus *Lagothrix* | woolly monkey | South America |

## SUBORDER ANTHROPOIDEA (Continued)

**Infraorder Catarrhina**
Superfamily Cercopithecoidea
FAMILY CERCOPITHECOIDAE
*Subfamily Colobinae*

| | | |
|---|---|---|
| Genus *Colobus* | guereza (colobus monkey) | Africa |
| Genus *Presbytis* | langur | India, Pakistan, Ceylon, Southeast Asia |
| Genus *Pygathrix* | douc langur | Laos, Vietnam |
| Genus *Rhinopithecus* | snub-nosed langur | West China, North Vietnam |
| Genus *Simias* | Pagai Island langur | islands off Sumatra |
| Genus *Nasalis* | proboscis monkey | Borneo |

*Subfamily Cercopithecinae*

| | | |
|---|---|---|
| Genus *Macaca* | macaque monkeys (includes Barbary ape and Celebes black ape) | North Africa, Asia, Southeast Asia |
| Genus *Papio* | common baboons, hamadryas baboon, mandrill, and drill | from sub-Saharan Africa to part of Middle East |
| Genus *Theropithecus* | gelada | Ethiopia |
| Genus *Cercopithecus* | guenon | Africa |
| Genus *Erthrocebus* | patas monkey | Africa |
| Genus *Cercocebus* | mangabey | Africa |

Superfamily Hominoidea
FAMILY HYLOBATIDAE

| | | |
|---|---|---|
| Genus *Hylobates* | gibbon | Southeast Asia |
| Genus *Symphalangus* | siamang gibbon | Sumatra, Malay Peninsula |

FAMILY PONGIDAE

| | | |
|---|---|---|
| Genus *Pongo* | orang-utan | Borneo, Sumatra |
| Genus *Pan* | chimpanzee | Africa |
| Genus *Gorilla* | gorilla | Africa |

FAMILY HOMINIDAE

| | | |
|---|---|---|
| Genus *Homo* | man | worldwide |

The above classification is presented for the student as a guide to the living primates. The author has borrowed from such prominent authorities as Dr. G. G. Simpson (who laid the foundation for modern primate taxonomy in 1945); Dr. W. Fiedler; Professor J. Buettner-Janusch; Dr. William L. Straus, Jr.; and, of course, Dr. Adolph H. Schultz. It is hoped the student will refer to this classification when dealing with the first part of the text.

# SUMMARY

Systematics, or Taxonomy
Archetype
Carolus Linnaeus
Bionomial Nomenclature
Genus
Species
Populations
Subspecies
Cline
Incipient Species
Law of Priority
Phylogeny
Parallel Evolution
Convergent Evolution
Natural Selection
Adaptive Trait
Preadaptive Traits
Nonadaptive Traits
Darwinian Fitness
Polytypic Species
Evolution
Irreversibility
Prosimians, or *Halbaffen*
*Anthropoidea*, or Simians
Lower Primates
Higher Primates
Platyrrhines
Catarrhines
*Ceboidea*
*Cercopithecoidea*
*Hominoidea*
*Hylobatidae*
*Pongidae*
*Hominidae*

Chapter 3

# The Treeshrew
## a living fossil

The treeshrews, or **Tupaiiformes**, present more problems to primatologists than any other living mammal. First, the treeshrew is misnamed. It is not a shrew nor is it exclusively an aboreal mammal. These little creatures live in bushes and the lower branches of trees in tropical rain forests and montane forests; one species *(Tupaia glis)* may be a ground dweller in certain locales. These are the least of the problems concerning the treeshrews. The big problem is whether these animals are in fact primates, insectivores, or something in between the two. We will briefly explore this issue.

### The Treeshrew Pattern

The treeshrews have a relatively wide distribution throughout Southeast Asia and some parts of Asia—India, Ceylon, Vietnam, Cambodia, Thailand, Malaya, Burma, Indonesia, the Philippines and minor islands. As mentioned above, these animals live in tropical rain forests and forests that border mountainous terrain. Generally the smaller the species

*Fig. 3–1* The common, or large, treeshrew (*Tupaia glis*). Note the lateral placement of the eye and the long muzzle with its "wet" rhinarium. (San Diego Zoo Photo.)

of treeshrew, the closer it lives to the ground. These creatures can at best be considered only **partially arboreal.**

At a glance, the treeshrew has a **rodent** or **squirrel appearance** (Fig. 3–1). In fact, the Malay word "tupai" means squirrel. These animals have both **long bodies and tails** with a dull coat color. **Little sexual dimorphism** is evident; males and females are almost identical in body size and form. The **muzzle is very long** and expands into an

**extensive rhinarium** or moist naked area like the dog's wet nose. The lips are **tethered**—that is, the upper lip is attached to the upper jaw (maxilla)—so facial expression is very limited. Unlike true primates, **all digits** are **clawed**. The treeshrew cannot climb or pick up objects by grasping. The **hands** are **not** truly **prehensile** although a convergent grip of both hands is employed, and there is some tendency towards independent movement of digits. The thumb is divergent from the other digits but is not opposable to them. The pen-tailed treeshrew is reputed to be somewhat aberrant in that its thumb and big toes are opposable, thus making its hands and feet prehensile.

Although some treeshrews can momentarily stand on their hindlimbs, they move about quadrupedally ("on all fours"). Their extremities are rather short for their bodies, with the hindlimbs somewhat longer than the forelimbs. Treeshrews are **diurnal** animals; their major activities are confined to the daylight hours except for the pen-tailed treeshrew, which appears to be a nocturnal creature. While most treeshrews have bushy tails, the pen-tailed variety has a tail that is bald except for a tuft of hair at the tip. It is believed the tails of treeshrews are used as balancing organs.

The treeshrew's **incisors** (front teeth) are primatelike in being not chisel shaped but simple, or **generalized**. The lower incisor teeth tend to be somewhat procumbent (directed horizontally) and are used to comb the body fur, thus they function much as the dental combs of some prosimians (see Chapter Four). They are not anatomically like the prosimians' dental comb. Like many prosimians, treeshrews also possess what is termed a sublingual organ, an extra tongue that is thought to serve as a kind of toothbrush which cleans the lower incisors.

These animals are **soft-food omnivores**; they eat fruits, vegetables, insects, and other small animals. A treeshrew may eat its own weight in food in a day's time.

Compared to other mammals of similar size, the treeshrew has a **relatively large brain**. This brain also shows some reduction in the olfactory area, and in the **visual** area some **expansion**. Yet even with these features, the treeshrew brain is largely like that of an insectivore. The visual expansion is further reflected in the **relatively large eyes** which are placed in the primitive lateral position, except in the case of the pen-tailed treeshrew, whose eyes are placed somewhat more forwardly. Unlike the generally accepted primates, treeshrews are considered to have **no stereoscopic vision**, but they may have color vision.

Unlike most primates, the treeshrews generally have multiple births—as many as three or four young at one time. Several species have a **primatelike placenta** and fetal membranes, although authorities do not agree here. **Multiple** pairs of **breasts** are also found; some species have three pairs. The treeshrew also has a **throat gland** which is likely used for scent marking its territory.

The above are only a few of the traits that make up the treeshrew pattern and should give the student some idea of the mixed nature of this animal.

### To Be or Not To Be . . . A Primate

That is the question, the question which a handful of primate biologists ponder. When G. G. Simpson (1945) gave us the now "classic" classification of mammals, the treeshrews were placed in the order *Primates*. Since that time a continuing debate has been waged as to whether the treeshrew is a primate, an insectivore, or something in between. Simpson was not the first to suggest that treeshrews have primate affinities. As early as 1872, T. H. Huxley had pointed out certain primatelike features of the treeshrews. In 1910 W. Kaudern, in 1922 A. Carlsson, and LeGros Clark in 1934 favored the inclusion of these creatures in our order; William King Gregory of the American Museum of Natural History also favored such. It was not until almost the second half of this century that the debate blossomed. The increase of interest in the treeshrews corresponds roughly to the recently rekindled interest in the primates.

Some mammalogists have placed the treeshrews, along with a group called the elephant shrews, in a separate order, the *Menotyphla*. This classification has not proved satisfactory because these two animals are probably not closely related. The treeshrews being a mixture of primitive mammalian and incipient prosimian characteristics, some authorities believe the treeshrew is best defined as an aberrant insectivore. Dr. William L. Straus, Jr., has reasonably proposed that the treeshrew be placed in a separate order of its own—the *Tupaioidea*. Giving the treeshrews primate status is attacked by some scholars with the argument that these creatures have diverged the least from the ancestral mammalian pattern while independently acquiring new traits—a combination of primitive retentions and possibly convergent evolution. The latter case is naturally difficult to prove. One of the better arguments against primate status is that given by Dr. W. C. Osman Hill, who feels their inclusion in the order *Primates* will bring about a "wholesale revision of the order," the *Primates* thus becoming undefinable. Those who do favor

**Table 3–1.**    The Treeshrew as a Primate

| Against | For |
| --- | --- |
| doesn't look like a primate | superficial resemblances to rodents and squirrels not reflected in internal anatomy |
| multiple breast pairs | number of breasts reduced from primitive mammalian pattern |
| muzzle too long | muzzle shorter than most insectivores; the baboon, a higher primate, has a massive muzzle |
| | middle-ear anatomy primatelike |
| | external ear small and humanlike |
| rhinarium too large and moist | some olfactory reduction |
| olfactory organs not replaced by vision to a great enough degree, primate visual features due to convergent evolution | significant expansion of visual area of the brain |
| no stereoscopic vision | retina of eye approaches basal primate structure |
| throat-chest scent glands | such glands found in some primates |
| nonprehensile hands and feet | digits resemble those of primates; thumb and big toes mobile |
| | tendency for independent function of digits |
| claws on all digits | |
| testes not posterior to penis | some accepted primates do not have the typical primate genitalia pattern |
| no baculum or *os clitoris* bones | man does not possess either of these |
| multiple births | upper limit on number of offspring |
| too short a gestation period | |
| some treeshrew species do not have a primatelike placenta | some do have such placentas |
| too short a maturation period | variable among accepted primates |
| many of the neurological features of the treeshrew are found in the primitive marsupials | relatively large brain; very large brain-body ratios; incipient temporal lobes of brain |
| | some serological (blood) tests close to primates |
| social behavior is unprimatelike | primate behavior is too variable to be stereotyped; some primates such as the aye-aye are aberrant in behavior |

their inclusion in our order can take some hope from the fact that once scholars were not inclined to even admit the lemurs into the order. On both sides of the issue are well-respected scholars who present rather convincing evidence for their viewpoints. Some of the technical arguments are briefly outlined in Table 3–1.

One of the most perplexing aspects of the treeshrew pattern is not anatomical but behavioral. Soon after birth, the treeshrew mother deserts her young; she does come back every two days, but only for brief feedings. None of the "tender" concepts of the primate mother-infant relationship, such as grooming and fondling, take place. One of the strongest cases for excluding the treeshrews from the *Primates* is this very unprimatelike behavior.

To the new student, this debate must seem like academic hair-splitting and rather silly. Yet this is really a very worthwhile debate. By continuous reexamination of this problem, not only are concepts redefined, but new information and ideas are brought to view. Tacit acceptance of opinions is not a hallmark of true science. While this author favors the aberrant insectivore concept, he favors even more the continuous examination of the problem.

### Treeshrews as Living Fossils

No matter what one's school of thought, there is agreement on one point—the first primates were probably very close anatomically to the living treeshrews. These first primates (about seventy million years ago) were probably ground-dwelling quadrupeds who were forced to become partially arboreal and eventually fully arboreal because of the successful and rapidly expanding rodent populations. The rodents were forcing the treeshrews out of their terrestrial econiches into new arboreal ones. The rodents have held their niches to this day and in fact, may be the most adaptively successful mammals on our planet. Some mammalogists view the primates and rodents as being separately derived from an ancient generalized insectivore. Some students rather convincingly argue that the rodents were in fact derived from the early primate stem.

Thus the treeshrew is a good "living fossil," or more correctly, a good **structural ancestor** for the primates. A structural ancestor need not have been an actual ancestor, but simply an animal whose total morphological pattern would have likely been possessed by the true ancestor. A structural ancestor can therefore be a fossilized animal or one actually living today. The treeshrew ideally fits this category.

## SUMMARY

TUPAIIFORMES

THE TREESHREW PATTERN

> Partially Arboreal
> Rodent or Squirrel Appearance
> Little Sexual Dimorphism
> Very Long Muzzle
> Extensive Rhinarium
> All Digits Clawed
> Hands Not Prehensile
> Diurnal
> Generalized Incisors
> Soft-food Omnivores
> Relatively Large Brain
> Visual Expansion
> Relatively Large Eyes
> No Stereoscopic Vision
> Primatelike Placenta
> Multiple Breasts
> Throat Gland

CONCEPT OF STRUCTURAL ANCESTOR

# Chapter 4

# The Prosimians

The **prosimians** or **lower primates** are not half-monkeys as the German word "halbaffen" implies. They are, as monkeys, apes, and man are, a **grade of primate organization**. The word "prosimian," like "monkey," is a very broad term for a part of the primate spectrum. If the student abandons the idea of primate stereotypes, he will find that he need not know the name or have knowledge of every primate species to understand the order. One should understand the several grades of primate organization. These grades are abstractions from nature: they are structural, physiological, biochemical, and behavorial. In this book, we are primarily concerned with the basic anatomical, or structural, aspects.

The suborder *Prosimii* is generally divided into three major groupings: the **Lemuriformes, Lorisiformes,** and **Tarsiiformes**. The lemurs are generally considered the most primitive and tarsiers the most advanced prosimians. As a group the prosimians are very primitive, yet they are highly evolved, or specialized. They have retained much of the primitive mammalian pattern, while in some cases they have also developed highly specialized traits. This is reflected in the prosimian pattern. There is no such creature as a typical prosimian. As pointed out by **G. G.** Simpson, this group shows more basic variety than any other primate group.

## The Prosimian Pattern

As in the case of the primate pattern, the patterns of the different primate groups reflect those of the order as a whole; and again no single trait characterizes the group, nor does every species possess each trait of the pattern.

The prosimians are **unusually varied** in appearance. Body lengths vary from a few inches to several feet. The entire group is a good example of what we call **mosaic evolution**. This concept holds that various

*Fig. 4–1*  Dental formulae. *A*—a hypothetical early generalized mammal $\frac{3.1.4.3.}{3.1.4.3.}$. It is presumed that primate dentition evolved from such a pattern. *B* and *C*—the dentition of a prosimian (Lemur). In the side view (*B*) the upper incisor teeth are seen to be much reduced, while the canine tooth is sharp and daggerlike. Behind the canine are three premolars and three molars. In the lower jaws the incisor teeth and the canine project forwards forming the "dental comb." The latter is shown from below (*C*). The front premolar tooth (*p*) has secondarily assumed the functions and shape of a canine. (*A*, after LeGros Clark, 1959; *B* and *C*, after LeGros Clark, 1965. Courtesy of the Trustees of the British Museum.)

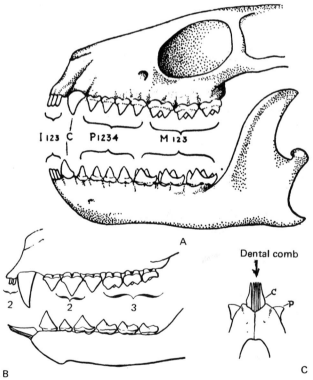

anatomical traits and organ systems have evolved at different rates and at different times—thus forming a mosaic pattern.

Most prosimians have a **nonprimate appearance**. Some look similar to raccoons, dogs, and even bears. Their **fur** often has **color designs**. Most have **tethered lips** in their upper jaws (as in the treeshrews), and therefore are capable of little or no facial expression. They also lack fully developed facial muscles. Above these lips is a moist **rhinarium** at the end of a **projecting muzzle**. Even though these animals have **relatively large eyes**, their cranial anatomy betrays a **great dependence on olfaction**.

The prosimian dentition exhibits both generalized and specialized features. The **dental formula** (D.F.) has been reduced from 38 teeth (in the treeshrew to $\frac{2.1.3.3.}{2.1.3.3.}$ (36 teeth in the typical prosimian (Fig. 4–1). There is some variation in this prosimian dental formula, thus it is a general pattern, not a rule of thumb. In diet the prosimians are generally soft-food omnivores. Frequently the lower incisor and canine teeth of prosimians form a procumbent **dental comb** which is used to groom their fur. As in the treeshrew, a sublingual organ (accessory tongue) is present and is used to clean the dental comb.

The hands of prosimians show considerable yet primitive dexterity in holding and manipulating objects. Their dexterity would be greater if it were not for the fact that their **digits act in unison**. Yet these prosimian hands are ideally constructed for maintaining powerful grips on tree trunks and branches. Individual digit mobility was likely selected against, while natural selection was requiring a very powerful grasp. A very peculiar specialization is the possession of a claw on the second toe of the prosimian foot; all other digits usually have nails. This structure is called the **toilet claw** (Fig. 4–2). As its name implies, it is used to clean the fur and skin of dirt and ectoparasites. It also can serve as a most effective back-scratcher. It is also called the grooming claw.

The prosimians are also unusual in that they have **multiple breast pairs**, yet generally only the pectoral pair are functional. The other ones serve as anchoring points for newborns.

There is **debatable evidence for color or stereoscopic vision** in this suborder. Unfortunately, differences in the color and pattern of the furry coat do not necessarily indicate that a group of animals possesses color vision, since these color differences can be interpreted by the noncolor vision eye as different shades of white-gray-black. The anatomical features of their retinas are interpreted differently by experts in this area. The **relatively large** and **forwardly placed prosimian eyes** are considered to be related to the high frequency of **nocturnal** activity found

T.C.

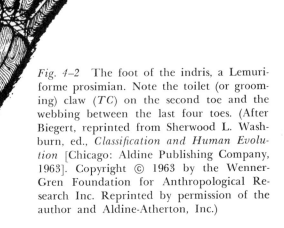

*Fig. 4–2* The foot of the indris, a Lemuriforme prosimian. Note the toilet (or grooming) claw (*TC*) on the second toe and the webbing between the last four toes. (After Biegert, reprinted from Sherwood L. Washburn, ed., *Classification and Human Evolution* [Chicago: Aldine Publishing Company, 1963]. Copyright © 1963 by the Wenner-Gren Foundation for Anthropological Research Inc. Reprinted by permission of the author and Aldine-Atherton, Inc.)

in the suborder. Some of these creatures also seem to be **crepuscular** in habit—that is, they are active at dawn and twilight, while a few may even be active at intervals throughout the twenty-four hour day.

While fossil prosimians lived in both the New and Old Worlds, today's prosimians are confined to sub-Saharan Africa, Madagascar, Southern India and Ceylon, and Southeast Asia (Fig. 4–3). It is of interest that while no nonhuman primates are found native to Australia, some of the marsupials show superficial similiarities to some prosimians—convergent evolution in similar niches. A good example is the case of the kangaroos and the prosimian hoppers, the galago and tarsier. There are even examples of similar dental and hand specializations in prosimians and marsupials.

As in the case of all nonhuman primates, quadrupedalism is the most common type of locomotor pattern in prosimians. In addition, a very novel pattern is found in many prosimians; we term this **vertical clinging and leaping.** These prosimians cling and rest in vertical positions and can leap considerable distances, coming to rest on a tree trunk again in a vertical position. (Fig. 4–4). If placed on the ground, these vertical clingers and leapers will run bipedally in some cases, and others will leap or hop. This locomotor adaptation is reflected in hindlimbs that are vastly longer than the forelimbs; these body proportions are suitable for arboreal (but generally not terrestrial) quadrupedalism. Their unusually long and bushy tails serve as balancing and propulsive organs in leaping.

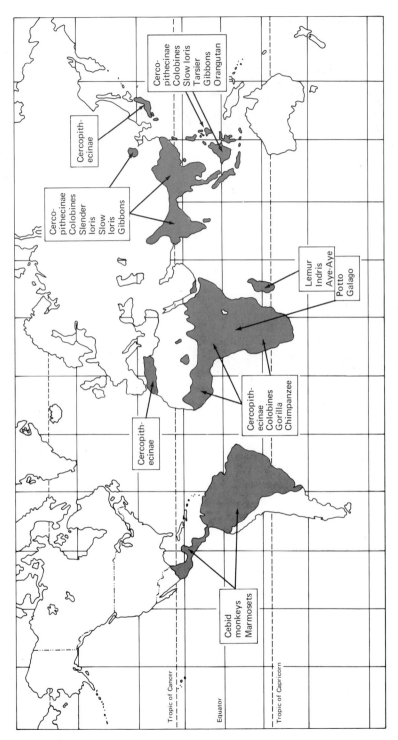

Fig. 4-3 Geographical distributions of the living primates. The African Lorisiformes are found throughout sub-Saharan Africa (potto and galago). The other Lorisiformes are in Southern Asia and mainland Southeast Asia. The gibbons are found throughout the mainland and islands of Southeast Asia.

The labels on the map read:

Cercopithecinae
Colobines
Slow loris
Tarsier
Gibbons
Orangutan

Cercopithecinae

Cercopithecinae
Colobines
Slender loris
Slow loris
Gibbons

Lemur
Indris
Aye-Aye
Potto
Galago

Cercopithecinae
Colobines
Gorilla
Chimpanzee

Cercopithecinae

Cebid monkeys
Marmosets

Tropic of Cancer

Equator

Tropic of Capricorn

*Fig. 4–4* Vertical clinging and leaping as seen in the indris and sifaka, two Lemuriformes. (After Napier and Walker, 1967, *Folia Primatologica*, S. Karger AG, Basel.)

## The Lemuriformes

The modern Lemuriformes are unique in that they are confined to the large island of Madagascar (Malagasy Republic) and the Comoro Islands in the Mozambique Channel between Madagascar and Africa (Fig. 4–3). How the lemurs got to this location is not understood. The idea that the Lemuriformes became isolated on Madagascar when the island broke away from a large supercontinent (due to continental drift) is not supported by the fossil record to date. One interesting idea is that the early Lemuriformes may have floated on beds of vegetation from the shores of Africa some thirty million years ago. We actually know more about why these creatures are not found elsewhere, in light of their once considerably broad world distribution. The cooling of the climates of later time epochs spelled extinction for these generally "temperature-fragile" creatures who require tropical climates. Competition from rodents, carnivores, and other primates (especially monkeys) eliminated the econiches available to the Lemuriformes. Madagascar is devoid of this type of competition for Lemuriformes; their only enemy now is man, who has drastically decreased their population in the last 100 years. With competition at a bare minimum, the Lemuriformes have successfully filled several econiches. All (except for the aye-aye) are vegetarians and are completely arboreal (except for the ring-tailed lemur). The conspicuous absense of animal protein from the diet of some and the factor of arboreality make for relatively exclusive econiches.

It is a misconception to think of the Lemuriformes as representative prosimians. This group has great variety—diurnal and nocturnal species, very large and very small body sizes, specialized diets, some very social, others quite solitary and unsociable. A few are full-time quadrupeds (dwarf and mouse lemurs), while others are vertical clingers and leapers. What these creatures are representative of are the econiches they fit so well.

### Types of Lemuriformes

Commonly the word "lemur" is used in place of Lemuriforme, and is derived from the Latin for "ghost." This name most likely came about in early descriptions of some lemurs who move rapidly and at night.

The lemurs are divided by this author into three main families or groups: the **common lemurs** (*Lemuridae*), the **indrises** (*Indriidae*), and the **aye-aye** (Daubentoniidae).

The lemurs have moved less away from the primitive mammalian pattern than any other primate group (except treeshrews if considered primates). Along with one or two other prosimians, they are considered to stand midway between the primitive insectivore ancestor and the monkey grade of primate organization. There are obvious advancements over the treeshrews such as the larger and more forwardly placed eyes, a trend carried even further by the lorises and the tarsier. Yet the lemurs are still rather primitive mammals. The females do not have the typical primate unicornate uterus, and multiple births are common. Maturation is very fast. Their brains are especially primitive, even when compared with the least evolved monkeys.

*Fig. 4–5* The ring-tailed lemur (*Lemur catta*). Note the relative lengths of the limbs. (San Diego Zoo Photo by F. D. Schmidt.)

COMMON LEMURS

The common lemurs are subdivided into two basic groups (subfamilies): the true lemurs, which are generally large and diurnal species; and the small lemurs, which is composed of the mouse and dwarf lemurs, both nocturnal creatures. The common lemurs are more quadrupeds than vertical clingers, yet some are quite agile leapers. This group exhibits considerable fur variation, a most attractive one being the ring-tailed lemur (Fig. 4–5). The ring-tailed species is rather unusual in terms of lemur locomotor patterns in that it is a terrestrial quadruped at times. Because of this ground-level movement and some behavioral traits, the ring-tailed lemur is sometimes called the "baboon" of lemurs. This is likely an overstatement of the facts.

The smallest living lemur is the mouse lemur, which weighs only a few ounces and whose tail is usually longer than its six-inch body (Fig. 4–6). Both the dwarf and mouse lemur are unique in their extreme susceptibility to temperature changes; this is related to their low basal metabolic rates. In cool weather these animals go into states of suspended animation, also termed torpor. During this period, the animal exists on the metabolic breakdown of fat deposits stored at the base of its tail. This is not a true hibernation. These tiny creatures are also unusual lemurs in their taste for animal protein, primarily insects; this is especially true of the mouse lemur.

THE INDRISES

The largest of the Lemuriformes are the indrises (Fig. 4–8). These primates reach dog-size proportions with body lengths, exclusive of tails, over three feet. Oddly enough, they are also somewhat doglike in appearance. Of all the Lemuriformes, the indrises are the most typically prosimian and primate. They are highly proficient clingers and leapers. This locomotor pattern is adaptably reflected in their very long hind-limbs and their preference for resting with their trunks erect in sitting positions. While being completely arboreal when placed on the ground, the indrises prefer to hop bipedally. In addition to the typical vegetarian Lemuriforme diet of fruits, flowers, and leaves, they also eat tree bark. There are three types of indrises: the common indris, the sifaka, and the avahi. All three have extremely well-adapted hands for grasping tree trunks and branches. All digits are long, especially the thumb, which is quite divergent (abducted) from the other digits for a prosimian. This pattern is usually seen also in the foot. The hands and feet of the indris and avahi are unique in that the second, third, fourth, and fifth

*Fig. 4–6*  The mouse lemur (*Microcebus maurinus*). Note the oversized eyes of this nocturnal primate. (San Diego Zoo Photo.)

*Fig. 4–7*   Erect posture in a lemur, the mongoose lemur (*Lemur mongoz*). Note the divergence of the big toe from the other toes and the balancing position of the long tail. (Courtesy of the Ministry of Information, Malagasy Republic.)

digits are partially webbed (see Fig. 4–2). Unlike the other Lemuriformes, the indrises do not have the typical prosimian dental formula. They lack a lower canine tooth and have lost one upper and one lower premolar tooth (D.F. $\overline{\frac{2.1.2.3.}{2.0.2.3.}}$). The avahi is the only nocturnal indris of the three types. The common indris has a short, stubby tail while the others have long ones. These primates have very thick fur. The largest type, the common indris, has a black and white coat and a dark head, while the medium-sized sifaka is pale white with a dark face. The smallest type, the avahi, is a gray-brown color.

*Fig. 4–8* The sifaka (*Propithecus verreauxi coquerelil*). Note the erect position of the trunk while at rest. (San Diego Zoo Photo by Ron Garrison.)

THE AYE-AYE

The most unusual of all the primate is a Lemuriforme called the aye-aye (Fig. 4–9). This animal is about the size of a small house cat. It has a very dark coat and a lighter-colored face. The aye-aye is a nocturnal primate and a very specialized one. This creature was once considered a rodent because of its very specialized dentition, which has some rodentlike traits (Fig. 4–10). Its adult dentition of 18 teeth (D.F.

*Fig. 4–9* The aye-aye (*Daubentonia madagascarensis*). Note the clawed digits, especially the unusual third finger. (Courtesy of the American Museum of Natural History.)

1.0.1.3.
1.0.0.3.) is the most unusual of all primates, even though its deciduous (milk) dentition is typically prosimian in number. Its incisor teeth are chisel shaped and have enamel only on one surface, giving a chisel-sharp edge to them. These incisors, especially the upper ones, have unusually long roots. Unlike the incisor teeth of other primates, the aye-aye's teeth grow continually throughout the animal's life. These incisors serve to strip bark from trees and also to open hard-shelled fruit.

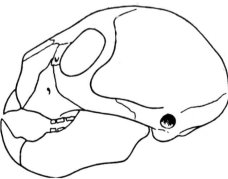

*Fig. 4–10* The skull and dentition of the aye-aye. Note the chisel-shaped incisor teeth and the similarly rodentlike jaws. (After LeGros Clark, 1959.)

Like its teeth, most of the aye-aye's anatomy reflects extreme adaptation to a highly prescribed niche. Claws are found on all of its digits except for the hallux (big toe). The middle finger of each hand is the most specialized digit of any primate. This finger is very thin and long and has a ghastly appearance. The Germans call the aye-aye the "fingertier," or the finger animal. This middle finger is used as an effective probe to skewer grubs (beetle larvae) that are in the wood beneath the tree bark. The bark is removed with the incisor teeth. Interestingly, one Australian marsupial which has adapted to the same econiche in that continent shows the same kinds of specializations as the aye-aye. A case has also been made for the aye-aye occupying the woodpecker's niche in Madagascar. Generally the elements of the aye-aye anatomy that are not involved in food collection are very primitive, especially its brain. For the biologist, the most highly evolved animal is the one that has adapted to its particular environment to the fullest; this means specialization. On the basis of this biological definition of specialization, one might argue that the aye-aye is the most evolved of all living primates. Since this animal is somewhat rare, there are many gaps in our knowledge of it. At present, the aye-aye is considered to be a vertical clinging quadruped.

The aye-aye is generally considered to be a relatively recent lemur adaptation. By ecological coincidence, this creature has adapted in a manner similar to that of the lemurs forty or more million years ago. Again we are presented with a living fossil.

The Lemuriformes are an excellent example of a group of primates that has undergone **adaptive radiation.** This is usually a very rapid adaptation to a number of specific econiches by a group of animals. In the process, populations undergo changes in anatomy and also increase their numbers significantly. Adaptive radiation is simple rapid speciation (and subspeciation) due to a lack of significant competition; and it is also usually due to the possession of a group of preadaptive traits. Perhaps as late as recent centuries, Madagascar lemurs as large as calves existed; they were ground dwellers and likely diurnal. Through his frequent unwise practices, man eliminated these giant lemurs from the ecosystem. Today we have almost exclusively arboreal, and predominately nocturnal, small lemurs. These lowest of primates must now be protected from the "highest."

While the behavior of the Lemuriformes does vary to a considerable extent, some generalizations can be carefully made. Approximately one half of the lemurs are vertical clingers and leapers and the other half quadrupeds of a branch-running and walking type. Their habitats being arboreal except in rare cases, their population sizes and ranges are difficult to measure. Group sizes seem to range from solitary individuals to perhaps twenty-five animals per group with the smaller lemurs being the more solitary. Groups tend to be larger at night; this is a general primate pattern. Their population densities, the number of animals per square unit of space, are fairly large; their quite prolific scent marking with glandular secretions and urine may be related to this. Some not only mark their general territories but also other lemurs. Dietary preferences are for leaves and fruit except in the notable case of the insectivorous aye-aye. Because of habitats and some nocturnalism, the behaviors of Lemuriformes are difficult to observe. In light of their endangered species status and interest in the "fossil behavior" of the earliest primates, further study of this primate group is imperative.

### The Lorisiformes

If one had to choose the most representative prosimians, it would be the Lorisiformes. These primates are wholly arboreal and almost completely nocturnal in habit. They are soft-food omnivores whose diets are made up of insects, small mammals, lizards, infant birds, bird's eggs, flowers, fruits, and vegetables. Obviously, the Lorisiformes are also predators.

In response to a nocturnal-arboreal niche, the Lorisiformes evolved large, forwardly placed eyes and quite powerful grasping hands and feet. Their thumbs and big toes are large and specialized, particularly in being placed almost directly opposite the other digits (Fig. 4–11). In superficial anatomy, the Lorisiformes are typically prosimian.

The Lorisiformes are divided into two families: the **lorises** and the **galagos**. These family groupings are largely based on locomotor patterns. The lorises are **slow climbers** and **creepers**, and the galagos are **fast hoppers**. These locomotor patterns have yielded important survival benefits to the Lorisiformes. While the Lemuriformes escaped competition from rodents, carnivores, and monkeys in the tropical forests of a large island, the Lorisiformes adapted to such competition in the tropical forests of Southern India, Ceylon, Southeast Asia, and sub-Saharan Africa. Their fitness is derived from their wholly arboreal nocturnalism, unusual locomotor patterns, and diets of mainly insects and small vertebrates. A fast hopper is very difficult to catch, while a slow creeper will be difficult to spot in the dark. Nocturnal activity also limits their competition for food and space. Like most mammals that are adapted to the night, the Lorisiformes have a special tissue layer (tapetum cellulosum) behind the retina. This layer acts somewhat like a combination mirror and magnifying glass, amplifying and reflecting faint light which enters the eyes at night.

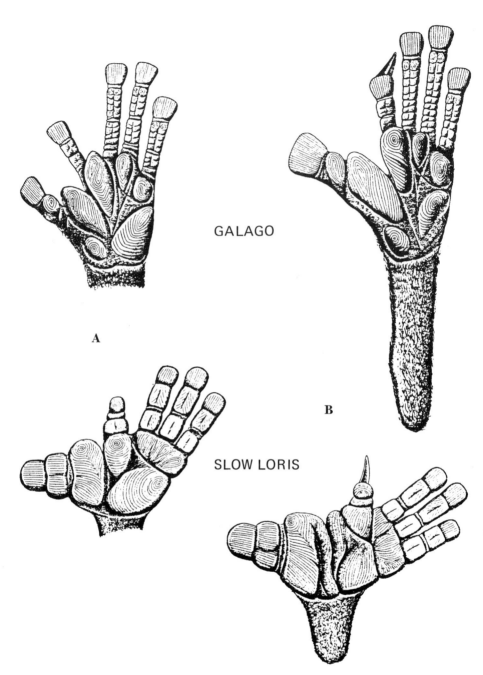

GALAGO

A

B

SLOW LORIS

*Fig. 4–11* The highly efficient grasping hands (*A*) and feet (*B*) of the Lorisiformes. (After Biegert, reprinted from Sherwood L. Washburn, ed., *Classification and Human Evolution* [Chicago: Aldine Publishing Company, 1963].
Copyright © 1963 by the Wenner-Gren Foundation for Anthropological Research, Inc. Reprinted by permission of the author and Aldine-Atherton, Inc.

*Fig. 4–12* The slender loris (*Loris tardigradus*). (San Diego Zoo Photo.)

## THE LORISES—SLOW CLIMBERS AND CREEPERS

The loris family consists of three rather similar animals: the slender loris, the slow loris, and the potto. The slender loris (Fig. 4–12) is found in Southern India and Ceylon and, like all the lorises, is a slow climber and creeper. Its body has been compared to a "banana on stilts." It lacks a tail. The slender loris maps out its territory by urine marking. Some of this marking is likely done when it goes through the procedure of washing its hands and feet with urine.

The slow loris (Fig. 4–13) is found in Southeast Asia. Its name well describes its movements. This creature is such a careful and deliberate climber and creeper that its movements seem almost ridiculous. The word "loris" is derived from a word meaning "clown." The name was probably derived from observation of its locomotor habits. It always has at least three extremities holding tightly to its climbing surface.

*Fig. 4–13* The slow loris (*Nycticebus coucang*). Note the efficient grasps of hands and feet. (San Diego Zoo Photo by Ron Garrison.)

*Fig. 4–14*   The potto (*Perodicticus potto*). Note the vertically slit pupils of the eyes and the almost absent index finger. (Courtesy of the Zoological Society of Philadelphia.)

Its movements are hand-over-hand and foot-over-foot. Like all Lorisi-formes, the slow loris can stand and sit erect.

While the slender loris has quite large external ears (a noctural adaptation), those of the slow loris are unimpressive. The eyes of this animal, while large, are not as voluminous as those of the other lorises. Typical of Lorisiformes, the pupils of the eyes form a vertical slit when contracted rather than the usual round primate form. The slow loris does engage in urine marking but has not been observed to urine wash.

The potto is a sub-Saharan African cousin of the slender loris. On first seeing this creature, one might take it for a small brownish bear (Fig. 4–14). The hands of the potto have the most perfect grasp of all living primates. The second finger is reduced to a mere tubercle, or stub, yielding ideal opposability of the thumb and remaining digits (Fig. 4–11). Due to a special vascular arrangement (rete mirabilia), the potto (and the two Asian lorises) is able to maintain a phenomenal grip. It is comparable to the body grasp of a boa constrictor. In captivity, the potto presents an excellent tug-of-war opponent for animal keepers.

Like the Asian lorises, the potto's limbs are almost equal in length; this is an adaptation to creeping quadrupedalism. The vertebral columns (backbones) of the Lorisiformes have the largest number of elements of any primates; this allows for more body flexibility in their locomotor pattern. The potto is unusual in that several of the cervical (neck) vertebral spines project through its skin. These exposed spines may serve to protect the sleeping potto and may also be used as offensive weapons. Typical of the lorises, the potto sometimes hangs and moves upside down, much in the manner of the South American sloth. The potto's body movements are so slow that in Africa it is called "softly softly."

Unlike the Asian lorises, the African potto is not known to practice any form of urine marking or washing. It does have anal scent glands which likely serve the same purpose.

There is a variety of the potto found in Africa that is sometimes called the golden potto, or angwantibo. This animal appears to be a smaller version of the potto. It exhibits differences from the common potto of a subspecies nature. To date little is known about this rare animal.

THE GALAGOS—FAST HOPPERS

The galagos, or bushbabies, are Lorisiformes but quite different from the lorises and the potto (Fig. 4–15). Like the potto, the galagos are found throughout sub-Saharan Africa but also on some offshore islands such as Zanzibar. The galagos exhibit great variety. They vary from the small dwarf bushbaby with its five-inch body to galagos that are house-cat size. All have long and very bushy tails. They also possess naked and mobile ears that aid their hearing in the dark. These creatures are almost, but not quite, wholly nocturnal. They are however completely arboreal and insectivorous in the wild. By reputation, they are known to make good house pets. Like the slender loris, the galagos engage in urine washing.

While the lorises and the potto may be termed vertical clingers and creeper-climbers, the galagos are the only Lorisiformes which are vertical clingers and leapers and quite excellent ones. They lack the excellent grasping hands of the potto. Their specialization is in their hindlimbs, which are grossly longer than their forelimbs. Two of the chief ankle bones (calcaneus and navicular) are greatly elongated to

*Fig. 4–15* The Senegal galago (*Galago senegalensis*). Note the external similarity to the tarsier. (San Diego Zoo Photo by F. D. Schmidt.)

*Fig. 4–16* The hopping locomotor pattern of the galago. (After Napier and Walker, 1967, *Folia Primatologica*, S. Karger AG, Basel.)

serve as a lever or spring mechanism in their fast hopping. Their kangaroo leaps or hops can carry them distances as great as seven feet along tree limbs and between trees (Fig. 4–16). When placed on the ground, these agile prosimians will also hop erect. A creature such as the galago may have been the ancestral stock from which the lorises and the potto arose.

*Fig. 4–17* The spectral tarsier (*Tarsius spectrum*). (Courtesy of the Oregon Regional Primate Research Center.)

### The Tarsiiformes

The tarsiiformes, or tarsiers, comprise the final and probably the most advanced group of prosimians in relation to the higher primates (Fig. 4–17). These small, specialized creatures live on islands of Southeast Asia. Creatures of the tropical rain forest, the tarsiers occupy in Southeast Asia approximately the same econiche the galagos do in Africa; they even look similar. The tarsiers are represented by only one genus *(Tarsius)*. The tarsier is an arboreal and nocturnal predator who prefers animal protein in the form of insects and small vertebrate animals. These are quite small, rat-sized primates that rarely exceed a half a pound in weight.

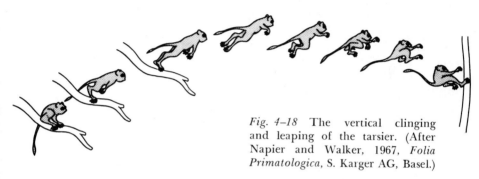

*Fig. 4–18* The vertical clinging and leaping of the tarsier. (After Napier and Walker, 1967, *Folia Primatologica*, S. Karger AG, Basel.)

Like the galago, the tarsier is an extreme vertical clinger and leaper, who moves in long leaps occasionally exceeding seven feet (Fig. 4–18). The tarsal bones of its ankle and foot are proportionately more elongated than those of the galago. The name "tarsier" is derived from these bones. To act as an anatomical shock absorber, the two leg bones (tibia and fibula) are fused in their lower portions; this is a primitive trait normally seen in quadrupeds. The tarsier's lower limbs are almost twice the length of its trunk. Its movements are most often compared to the hops and leaps of a frog. The tarsier also has a very long tail which is generally naked except for hair tufting at its end. The underside of this tail has dermal ridges like those found on our hands and feet. In spite of these "tailprints," the tarsier's tail is not prehensile like New World monkey's tails. The tail is used as a balancing organ in movements and also as a sort of third leg of a tripod when resting. The tarsier prefers an erect posture at all times.

Like the aye-aye and the galago, the tarsier depends greatly upon vision rather than a good sense of smell. This is reflected in the very advanced anatomy of the tarsier skull (Fig. 4–19). The muzzle has been reduced to a mere nubbin, or true primate nose, lacking the primitive

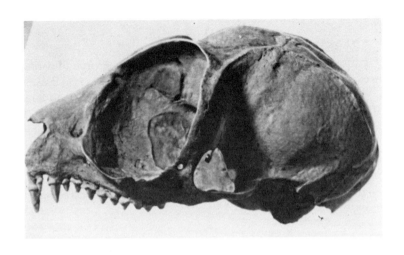

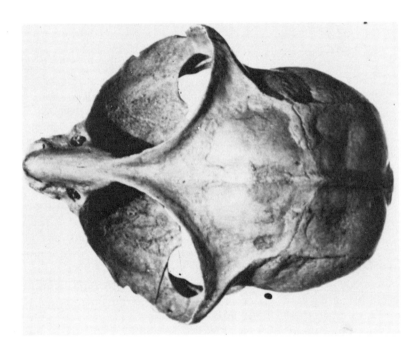

Fig. 4–19 The skull of the tarsier. Note the massive eye orbits and short muzzle. (Courtesy of the American Museum of Natural History.)

rhinarium. The only other prosimians to have undergone almost as much snout reduction are the small galagos, and this may be a function of their body size. The eyes of the tarsier are simply enormous. In volume the capacity of the bony eye orbits (sockets) exceeds that of the brain case. Primatologists often note that the tarsier is one animal whose eyes are actually larger than its stomach; this is true. While all prosimians have what is called a postorbital bar of bone to the sides of their eye sockets, only the tarsier has a postorbital plate behind each eye. These two anatomical features (especially the plate) keep the eyeballs from being pressed against by the powerful temporal muscles to their sides. This setup of bar and plate potentially allows for eye movement. The tarsier is unique in that it has both the proper complement of eye muscles for movement and this bony setup, yet these muscles remain vestigial (functionless). The tarsier compensates for this, as the more primitive prosimians do, by head movements. The joint between its skull base and the spine is so constructed to allow the tarsier head movement in approximately a 180-degree arc, much like an owl's. The skull base also shows an important advance over the other prosimians. The skull-base joint and the foramen magnum ("large hole"), which transmits the spinal cord, are more forwardly placed than in any other prosimian. This is likely related to the tarsier's habitual erect posture. We shall deal with this important anatomical area in evaluating fossil candidates for man's ancestry.

The tarsier has a free and mobile upper lip which lacks a cleft. Its face also has musculature enabling it to make "faces." The tarsier also lacks the dental comb but is aberrant in lacking one set of lower incisor teeth $\frac{2.1.3.3.}{(1.1.3.3.)}$.

Its hands have pads like suction cups on the ends of the digits. These pads enable the tarsier to move up vertical surfaces easily, even glass. Most significant is the relatively large tarsier brain. It shows a significant reduction in the olfactory area and increased development in the visual area. The tarsier is not an exaggerated galago; it is a quite unique and advanced prosimian.

A few anthropologists and anatomists have been so impressed with certain advanced features of the tarsier such as snout reduction, brain size and conformation, and the eyes that they have given this creature unusual status. The English anatomist F. Wood Jones (1923) claimed man was descended directly from the tarsier. Today Dr. W. C. Osman Hill places the tarsiers with the *Anthropoidea* in a separate suborder (the *Haplorrhine*). Many students of the primates point to the monkeylike features of the tarsiers and accord them undefined status between the New World monkeys and the prosimians, something like "half-monkeys."

The tarsiers are probably a unique experiment in primate evolution that foreshadowed higher primate trends. Due to some parallel development, the tarsiers acquired some higher primate traits. The living tarsiers are quite specialized as prosimians. They have two toilet claws (second and third toes). They have a prosimian-type uterus and a higher primate placenta. Their eye orbits are exaggerated and are not those of a higher primate. Specialized in some features and unusual in others, tarsiers are best viewed as aberrant primates that have survived into the present.

As you have seen, the prosimians are so varied that one can make few generalizations about them. By all rights, time should have passed them by; but through adaptation to precise econiches, they have survived. This long-term survival is also due to the general plasticity which the primate pattern provides. It is very likely that many of the features of the primate pattern, especially single births, were derived from adaptations of arboreally active and nocturnal creatures such as these. As long as we maintain and conserve these quite distant relatives, an understanding of the primate past is within our reach.

## SUMMARY

GRADE OF PRIMATE ORGANIZATION
LEMURIFORMES
LORISIFORMES
TARSIIFORMES
MOSAIC EVOLUTION
THE PROSIMIAN PATTERN
     Unusually Varied
     Nonprimate Appearance
     Fur Color Designs
     Tethered Lips
     Rhinarium
     Projecting Muzzle
     Relatively Large Eyes
     Great Dependence on Olfaction

Dental Formula $\dfrac{2.1.3.3.}{2.1.3.3.}$

Dental Comb
Digits Act in Unison
Toilet Claw
Multiple Breast Pairs
Debatable Evidence of Color and Stereoscopic Vision
Relatively Large and Forwardly Placed Eyes
Nocturnal
Crepuscular
Vertical Clinging and Leaping

COMMON LEMURS

INDRISES

AYE-AYE

ADAPTIVE RADIATION

LORISES

SLOW CLIMBERS AND CREEPERS

FAST HOPPERS

TARSIERS

Chapter 5

# The New World Primates

The monkey grade of primate organization is extremely difficult to define. The most common definition of the word "monkey" is "a tailed anthropoid." Compared to the prosimians, monkeys approach man and the anthropoid apes significantly closer in anatomy, biochemistry, physiology, and social behavior. Superficially the Old World monkeys and many of the New World monkeys look very similar to the nonprofessional eye. These surface similarities are probably due to a parallel evolutionary development from a very distant ancestor, at least sixty million years ago. It is generally held that this common ancestor was a tarsierlike prosimian: a protoprimate insectivore is also a possibility. The New World monkeys are thought to have originated from a fossil prosimian family called the omomyids, which were indigenous to North America.

The New World primates are called the **Ceboidea** or *platyrrhines* because of their distinct nose shapes (Fig. 5–1). To call all of these creatures "monkeys" is stretching a point somewhat because some of the most simple of these primates (marmosets) are virtually monkeys only by possession of tails. I prefer to call them collectively New World primates.

For simplicity the New World primates are divided into two families: the **marmosets** and the **cebid monkeys**. These primates inhabit the tropical forests of Central and South America and extend even into parts of Southern Mexico (see Fig. 4–3).

More precisely, we can divide the New World primates by locomotor pattern, in this case borrowing the terminology of the distinguished American primate anatomist, Dr. George E. Erickson: marmosets are **springers**—this is their main means of locomotion; and cebid monkeys are **springers, climbers,** and **brachiators.**

### The New World Pattern

The New World primates have **platyrrhine noses** (platy = "flat," and rhine = "nose"). Round nostrils, widely separated by a broad fleshy nasal septum, give the New World nose a very flat appearance. This nose type is distinct from the Old World anthropoid, or catarrhine, nose (Fig. 5–1).

*Fig. 5–1* A New World primate (*A*) and an Old World monkey (*B*). Note the the shapes of the nostrils and the septa between them. (After LeGros Clark, 1965, Courtesy of the Trustees of the British Museum.)

The most striking feature of the New World pattern is the **great variety** of types, body forms, and sizes; this variety is also reflected in their internal anatomies. Creatures of a few ounces and inches are found, as well as medium dog-sized ones. Complex evidence beyond the scope of this text points to the New World primates being much closer genetically to the prosimians than the Old World monkeys. Even with this lack of anatomical uniformity, there is **little sexual dimorphism** among the New World primates compared to their Old World counterparts.

**All** New World species **have tails** with **some** being **prehensile**— that is, capable of grasping. This is particularly indicative of the fact

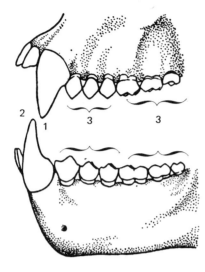

*Fig.* 5–2 The New World dentition (*Cebus*). (After LeGros Clark, 1959.)

that the New World primates are **completely arboreal**. Living in such an environment, there is no need for a large, aggressive male to protect the females and young against ground predators; this may explain their only minor sexual dimorphism. Under natural conditions, they rarely come to the ground. Water is obtained from their food and by collecting rain water on vegetation. This arboreal adaptation is partially reflected in their **long limbs** whose digits possess **long curved nails**, which are technically modified claws. The **mobility of** these **digits is less** than that found in the Old World monkeys. Interestingly, many of the New World primates have nonopposable thumbs; yet their big toes are pseudo-opposable and have considerable capabilities. In spite of their arboreal habitat, these primates **lack** the **ischial callosities**, or sitting pads, found in Old World monkeys (see Chapter Six).

The New World primates retain an essentially prosimian **dental formula** of $\frac{2.1.3.3.}{2.1.3.3.}$ but without the dental comb (Fig 5–2). A few species of marmosets consistently lack third molar teeth, while many other New World primates show population trends toward reduction in size of the tooth and loss of these "wisdom" teeth. The significance of this is not clear. The New World primates also **lack** the **cheek pouches** that many Old World monkeys possess for storage of food.

Female New World primates, unlike some of their Old World counterparts, have **no sexual swellings**. These external swellings occur among Old World monkeys when ovulation is near and the female is sexually receptive. Behaviorally they are somewhat unique in that male marmosets care for the infant offspring. The owl monkey, another New World species, also exhibits this behavior.

**The Marmosets**

The marmosets are quadrupedal runners and leapers and are classified as springers; their movements are comparable to those of a squirrel.

*Fig. 5–3*   Goeldi's marmoset (*Callimico goeldii*). Male at top, female carrying its baby. (San Diego Zoo Photo by Ron Garrison.)

These South American primates are very small and rather uniform in their anatomy for New World primates. The smallest New World primate is the tiny pygmy marmoset, which can easily be held in one's hand. The marmoset's coat is generally silky and dense (Fig. 5–3); one group of marmosets, the tamarins, exhibits a variety of head-hair tufts and facial moustaches. The marmoset tail is long and acts as a balancing organ in movements and as a support in erect resting postures.

The marmoset brain is very primitive for a primate classified as a monkey. These small creatures seem to rely on their sense of smell to a great extent because they lack the visual acuity of other monkeys. Like prosimians, they possess anal scent glands. Multiple births, usually twins, are common among them. Whereas we saw the retention of one claw (the grooming claw) in the prosimians, the marmosets have long, modified claws on all digits except the big toe, which has a flat nail. These claws enable them to climb trees like squirrels rather than like primates.

In diet, marmosets prefer insects, fruit, vegetables, lizards, birds and their eggs. The tamarins are unique in that they prefer to be active predators rather than food gatherers. As noted before, the marmoset dentition is unusual in its usual lack of a third molar— D. F. $\frac{2.1.3.2.}{2.1.3.2.}$. The jaws are shorter in response to this loss. Goeldi's tamarin (Fig. 5–4) is unique in having retained the small third molar. Furthermore, this marmoset shows some anatomical traits similar to tarsiers and others similar to the cebid monkeys; some authors classify it as a cebid monkey. As with other marmosets, Goeldi's tamarin to a certain degree represents a good living structural ancestor that fills the gap between prosimians and true monkeys as well as between marmosets and cebid monkeys. The primitive features of the marmosets make the collective terminology New World "primates" preferable to New World "monkeys."

The behavior of marmosets is not typically primatelike. They are one of the few primates that live in monogamous units, male and family; the owl monkey, titi, gibbon, and man are the only other nuclear family primates. The marmosets infant-adult relationship is unusual—the male, not the female, carries and tends to the infants, except at feeding time. This caretaking can be quite a task because of the common occurence of twins. This situation may be related to their monogamous family setup and possibly to their lack of sexual dimorphism. These families live alone or in groups ranging up to eight members. Distance communication is by stereotyped calls, which have been described as birdlike.

*Fig. 5–4* The golden lion tamarin (*Leontideus rosalia*). (San Diego Zoo Photo.)

## The Cebid Monkeys

### SPRINGERS

The subfamily *Aotinae* is unusual for a cebid monkey group in being both marmoset—and prosimianlike in appearance. They are also unusual in being springers in locomotor pattern. This group is the most primitive of the cebid monkey family and is made up of two animal types, the titi (Fig. 5–5) and the owl (or night) monkey (Fig. 5–6). The owl monkey is also known as the dourocouli. These monkeys are small and medium sized; both types are omnivorous. Of the two, the titi appears more like a marmoset. The titi has large eyes yet is diurnal in habit. The owl monkey is the only nocturnal New World primate;

*Fig. 5–5* The widow titi (*Callicebus torquatus*). (San Diego Zoo Photo by Ron Garrison.)

*Fig. 5–6* The dourocouli, or owl monkey. Note the eye sizes (San Diego Zoo Photo by Ron Garrison.)

it has very large eyes. The owl monkey is further unique in that it is in some ways a vertical clinger and leaper as well as a springer. It also lacks independent digit mobility as is the case in prosimians. Some authorities consider the owl monkey a likely structural ancestor between the prosimian and monkey grades of primate organization. As with all living primates, the evolutionary relationships are not clear.

Ecologically the owl monkey appears to fit a nocturnal, prosimian niche in the New World. In fact, a good part of its diet is composed of insects, as are the diets of the Lorisiformes and the tarsier. The owl monkey also washes in urine to mark its territory. Living in groups up to six in number like the titi, its basic type of social organization is the "family."

The titi is one of three New World primates that occupies parallel aspects of the behavioral niche that the gibbon, an anthropoid ape, holds in Southeast Asia (the other two are the howler and spider monkeys). The titis have high population densities; their home ranges are small, so small that their territories and home ranges are virtually the same. In response to this type of arboreal setting, they have evolved rich vocal repertoires surprisingly like those of gibbons, whose territories are similar. Like the gibbon, the titis' vocal displays occur mostly in the morning and are used to maintain social distance between groups. In addition to such vocal marking, they also engage in scent marking.

### CLIMBERS

The first subfamily of climbers is the *Pitheciinae.* This group is composed of the saki and uakari monkeys, two very exotic looking creatures. Unlike other New World species, these monkeys possess a dental arrangement which can best be described as a pseudodental comb. Sakis can be either bearded or plain-faced (Fig. 5–7). Both types have very long, bushy tails and resemble to some degree another New World primate, the howler monkey. The sakis are unique in that they do show marked sexual dimorphism.

The uakaris are probably the most unusual looking of all primates (Fig. 5–8). Bald-headed, with bare, hourglass-shaped faces, shaggy-coated, and stub-tailed, these vegetarians present a sight unduplicated elsewhere in our order. One species, the red variety, has a reddish face that is said to change hue with its moods.

The other New World climbers are in the subfamily *Cebinae,* which is made up of two animals that are probably more familiar. They are the squirrel monkey and the capuchin (organ grinder's) monkey. Both are found in Central and South America. Both monkeys show a tendency toward reduction in size of and loss of the third molar tooth.

The squirrel monkey is as its name implies, a squirrel-sized creature (Fig. 5–9). With its large ears and short, dense coat, it is readily identifiable by the white "goggles" around its eyes and its black snout.

*Fig. 5–7* The monk saki (*Pi-thecia monachus*). (San Diego Zoo Photo by Ron Garrison.)

*Fig. 5–8* The red uakari *(Ca-cajo rubicundus)*. (San Diego Zoo Photo by Ron Garrison.)

The skull of this nonhuman primate is the most humanlike of all New World primates. In fact relative to body size, its brain is proportionately larger than ours.

*Fig. 5–9* The common squirrel monkey (*Saimiri sciureus*). (San Diego Zoo Photo by F. D. Schmidt.)

Like the owl monkey, the squirrel monkey eats a lot of insects and also marks its territory by urine washing. To a certain extent, this primate seems to be a behavioral parallel in the New World of the African talapoin (*Cercopithecus talapoin*); both forage in small groups but form troops up to 100 members as sleeping groups. Male dominance behavior in the squirrel monkey, as evidenced in the semicaptive situation, indicates that males may be truly dominant only in the mating season, and are ostracized from the group the rest of the time.

*Fig. 5–10* A tufted capuchin monkey (*Cebus apella*) and a four-month-old infant. (Courtesy of the Zoological Society of London.)

The capuchin monkeys are of two types, one with and the other without tufts of hair on the head (Fig. 5–10). These capuchins are quite lively and extremely intelligent primates; they can also be rather nasty tempered at times. They have prehensile tails, yet these tails do not have a specialized bare-skin area as do those of the brachiating New World primates. It is not unusual for the tails of infant monkeys to have some prehensibility; this is usually lost with maturation. Schultz has pointed out that the squirrel and capuchin monkeys parallel to some extent the body conformations and life styles of some Old World monkeys such as guenons and macaques.

Like the squirrel monkey, the capuchin also marks its territory by urine washing; it also uses chest gland secretions in marking behavior. For a long time, investigators have reported the great manual dexterity of the capuchin monkey. In recent years a new dimension has been found in its behavior—tool-using and possible tool-making. In the laboratory capuchins have used sticks as weapons and in the wild threaten predators and man by dropping tree branches and other arboreal objects to the ground. We now know they also modify branches and leaves to probe for larvae; one would define this as tool modification for an immediate purpose. For years man was defined as the "tool-maker." The discovery by Jane Goodall that chimpanzees in the wild also modify naturally occurring objects to use as tools has greatly changed our concept of tool-making and man's uniqueness. The capuchin monkey is adding to the evidence that "tool-making" may not be an isolated behavioral phenomenon in nonhuman primates but a general pattern.

### BRACHIATORS

Brachiation is a hand-over-hand, arm-swinging locomotor pattern in which the body is suspended by the upper extremities. As children we all attempted brachiation on "monkey bars" and "jungle gyms." This locomotor pattern is also a function of feeding behavior. The brachiating primate, usually a large animal, is able to obtain food from tree branches which normally would not support its weight in a quadrupedal position. By suspending himself from such tree limbs, the brachiator is able to safely reach fruit and leaves on slender branches.

The New World brachiators possess prehensile tails which serve as "third hands" or "fifth limbs." The end third of this tail is naked on the underside. This bare patch of skin has dermal ridges akin to those found on our hands and feet. These "tailprints," along with very strong tail muscles, give this tail an amazingly strong and accurate grasp. Dr. John Napier prefers to call the New World brachiators "semi-brachiators," while some prefer the term "arm-swinging quadrupeds." They are best termed brachiators, for this is what they exhibit—one of the several forms of brachiation. The New World brachiator will often suspend himself by his tail and swing or drop to another branch; the tail not only serves as another hand but also frees the true hands for grasping other branches and food. Regular hand-over-hand arm-swinging is also exhibited. As in the case of the Old World brachiators (who do

Howler                                    Woolly

*Fig. 5–11* The comparative limb and trunk proportions of the New World brachiators. (Courtesy of G. E. Erickson and the Zoological Society of London.)

Woolly Spider                             Spider

not have prehensile tails), the New World brachiators show a trend toward short trunks and long limbs, especially the forelimbs (Fig. 5–11). The New World brachiators are divided into two subfamilies: the Alouattinae, consisting of the howler monkeys; and the Atelinae—woolly monkey, spider monkey, and woolly spider monkey.

Like most of the New World brachiators, the howler monkey (Fig. 5–12) moves mostly in quadrupedal gaits rather than brachiation. This monkey is actually a **structural brachiator** rather than a functional one. The howler has a brachiator's anatomy but seldom really brachiates. This creature grasps between its index and middle fingers rather than the thumb and index finger. This grasping pattern is found to some degree in many New World primates and is foreshadowed in some prosimian manual activities. This is actually a primitive mammalian pattern and is also exhibited in some marsupials.

The howler monkey gets its name from the loud calls it makes. It has a very specialized vocal apparatus, especially its hyoid bone. In most primates, including man, this bone is a very small structure. In the howler the hyoid is huge and inflated like an eggshell to serve as

*Fig. 5–12*  The red howler monkey (*Alouatta seniculus*). (San Diego Zoo Photo.)

a vocal resonating chamber. Much of the howler's cranial anatomy is fashioned to support this large vocal apparatus.

Primarily a leaf-eater, the howler also eats other vegetation, especially fruit, which it can eat in the unripe stage. The howler has evolved, like some Old World colobine monkeys, the unusual ability to digest mature leaves; a complicated digestive tract is required for this behavior. The howler exists in a number of species. The most spectacular of these are the black and the red howlers.

The behavior of the howler monkey has drawn the attention of students of primate behavior for a long time. Organized into groups varying from several to fifteen animals, the howlers have rather fluid home ranges. It is not unusual for the ranges of groups to overlap. The howler's territory might be termed an "auditory territory" because groups do not usually make visual contact but vocal contact. Vocalizations are most prominent upon waking in the morning and when beginning movements in the afternoon. Males use vocalization to initiate group movement. Social distance between groups is maintained by vocalization, thus we have "avoidance" vocalization. The howlers are unique among primates in that they repulse other groups by vocalization alone without any physical or gestural threats. Being unaggressive in behavior, male howlers do not clearly exhibit dominance. Leadership in howling and initating it may be the realm of dominance for the male. A male howler will drop branches and empty it bowels and bladder on human observers below.

Another prehensile-tailed New World brachiator is the woolly monkey (Fig. 5–13). It is named for its dense, woolly fur. This dark-colored primate is a large, robust monkey who possesses an equally robust tail. "Pot-bellied" and quiet, the woolly monkey is a very docile and friendly animal that can make a good pet.

*Fig. 5–13* The woolly monkey (*Lagothrix lagothricha*). (San Diego Zoo Photo.)

The spider monkey is the most thoroughly adapted to brachiation of all the New World brachiators (Fig. 5–14). For good brachiation ability, the hand must be able to form an effective **hook grip**. However, the thumb of many primates, including man, gets in the way. Selective evolutionary pressures for a brachiating mode of locomation can bring about a number of anatomical solutions to the problem of obtaining this hook grip: (1) bringing the thumb into alignment with the other fingers, as is the case in the howler and woolly monkeys, (2) lengthening the fingers, (3) placing the thumb back towards the wrist, (4) reducing the thumb in size to where in some cases it is either a mere nubbin (tubercle) or is totally absent, (5) possibly enabling the thumb to be placed across the palm or atop adjacent digits in brachiation, or (6) combining all of these features to varying degrees. The woolly and howler-monkey fingers have been aligned and somewhat lengthened. The spider monkey's thumb is usually just a tubercle and quite often is absent. Like the other New World brachiators, if placed on the ground, the spider monkey will often prefer bipedal running to quadrupedalism; some authorities consider the brachiator's anatomy as a preadaptation for erect posture and bipedal locomotion.

In its almost perfect adaptation to its New World arboreal niche, the spider monkey occupies the same econiche that the gibbon holds in Southeast Asia; the gibbon is considered to be the most efficient of all brachiating anthropoid apes. The spider monkey shows locomotor and anatomical features similar to the gibbon. Both the gibbon and spider monkey are quadrupedal at times, so there is no such creature as a "full-time brachiator."

Spider monkeys make acceptable pets; but like most New World species, they are very susceptible to infectious diseases and do not live long in captivity.

A very rare New World brachiator is the so-called woolly spider monkey. This prehensile-tailed species has the coat and body size of the woolly monkey, but in other features, especially the lack of a thumb, it more closely resembles the spider monkey.

The New World primates are indeed a varied lot. The living representatives of this group are probably the result of a very early and rapid adaptive radiation to the tropical forests of the New World. Their adaptations to these arboreal environments are generally extreme when compared to their Old World counterparts. Like the prosimians, the New World primates are a unique experiment in primate evolution, an experiment from which we are fortunate enough to have the living results.

*Fig. 5–14* The black-handed spider monkey (*Ateles geoffroyi*). Note the hook grip of the free hand and the lack of a thumb. (San Diego Zoo Photo.)

## SUMMARY

CEBOIDEA

PLATYRRHINES

MARMOSETS

CEBID MONKEYS

SPRINGERS

CLIMBERS

BRACHIATORS

THE NEW WORLD PRIMATE PATTERN

Platyrrhine Noses
Great Variety
Little Sexual Dimorphism
All Have Tails—Some Prehensile
Completely Arboreal
Long Limbs
Long, Curved Nails
Less Mobility of Digits
Lack Ischial Callosities

Dental Formula $\dfrac{2.\ 1.\ 3.\ 3.}{2.\ 1.\ 3.\ 3.}$

Lack Cheek Pouches
No Female Sexual Swellings
Structural Brachiator
Hook Grip

# Chapter 6

# The Old World Monkeys

All of the Old World *Anthropoidea* (monkeys, apes, and man) are sometimes called **catarrhines**, referring to their particular type of nose conformation. To the primate biologist, the catarrhines are closer to one another biologically than they are to the New World primates or prosimians.

The Old World (catarrhine) monkeys are placed under the broad superfamily *Cercopithecoidea* (family Cercopithecidea). There is a further division into two large subfamilies: the *Colobinae*, or **leaf-eaters**, and the *Cercopithecinae*, or **cheek-pouched monkeys.** Names such as "leaf-eater" are convenient racks for complex coats. Perhaps a more functional terminology for Old World monkeys would be based on habitat and adaptation to it. One can subdivide these monkeys first into **forest-dwelling browsers** who leisurely eat fruit, leaves, birds, and other forest life. These browsers, who are mainly arboreal, react to danger from predators by "flight." The other functional category is **grass-eating savannah dwellers.** These monkeys usually inhabit the edges of savannah grasslands and savannah woodlands and exist mainly on an herbivorous diet of grass stalks and roots. These grasses contain significant amounts of silica (sand), bringing considerable wear to molar teeth and also requiring a special dental apparatus. The savannah dweller reacts to danger by "fight"—that is, aggressive or decoy behavior. Dominant males will defend, often by decoy, weaker members of the group, who are thus able to flee the danger of a predator. An actual fight is rare. The colobines are basically forest-dwelling browsers, while the Cercopithecinae are made up of both types.

## The Old World Monkey Pattern

The Old World monkeys are more **widely distributed** than the New World primates or any other group of nonhuman primates. These Old World species are found from the Rock of Gibraltar to the islands of Southeast Asia and from the tip of South Africa to the Himalayas, and even to the most northerly reaches of Japan. With such a vast geographic distribution one might expect a similar diversity in morphology. Surprisingly the Old World monkeys show a remarkable **comparative uniformity of anatomy** when compared to their New World counterparts. The work of Adolph Schultz has eloquently proven this to be the case.

Like the anthropoid apes and man, the Old World monkeys have **catarrhine** ("down-pointed") **noses**. The nose conformation is two comma-shaped nostrils separated by a rather thin and often tapering, fleshy nasal septum. These monkeys also share with the apes and man the dental formula of $\frac{2.1.2.3.}{2.1.2.3.}$ (Fig. 6–1).

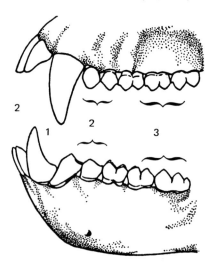

*Fig. 6–1* The catarrhine dentition as seen in a macaque monkey. (After LeGros Clark, 1959.)

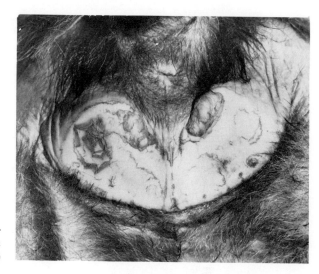

*Fig. 6–2* The ischial callosities of a young male baboon (*Papio anubis*). (S. I. Rosen.)

As with the New World primates, most of these monkeys have some type of tail, but **no prehensile tails** are found. Nocturnalism has not been documented for these monkeys, thus **all** are thought to be **diurnal**. Unlike the gracile New World primates, the Old World monkeys tend to be **robust**. Their vocalizations are also much deeper than the shrill voices of many of the New World primates.

The Old World monkeys are unique in possessing **ischial callosities,** or so-called sitting pads (Fig. 6–2). These hard, callused skin areas are located on the buttocks and are even present before birth. They serve as platforms for sleeping and resting on tree limbs (Fig. 6–3). They are also useful to terrestrial Old World species such as the baboon, who often stops and sits while eating. The ischial callosities may in addition serve to protect the lower limbs from having their blood supply interrupted and nerve supplies crushed during prolonged sitting. Similarly unique is the **varity of sexual swellings** exhibited by some of the Old World monkeys. When in oestrus, the time of ovulation and sexual activity for the female, her sexual skin and occasionally also the external genitalia become swollen and attain a deep reddish color. This female sexual swelling serves as a visual cue to males that the female is sexually receptive at that time; this may not hold true for all Old World species, since it is not clear how large a role sexual odor also plays in the mating behavior. The females of some Old World species exhibit a much lesser swelling and reddening of the sexual skin (and sometimes lower abdomen). The results of their sexual activity is usually **single births.**

*Fig. 6–3* An adolescent female proboscis monkey (*Nasal larvatus*) resting on its ischial callosities in a tree. (Courtesy of the Sarawak Museum.)

The Old World monkeys are **biologically closer to man** than the New World primates. This has made a number of these species, especially the rhesus monkey and the baboon, choice animals for biomedical research.

**Sexual dimorphism** is **quite common** in these monkeys, whereas it is rare among New World species. The most exaggerated sexual dimorphism is found among the terrestrial savannah dwellers such as the baboon and the patas monkey. Considerable body-size differences (males sometimes being twice the size of females), prominent frontal fur, and large, projecting canine teeth are generally conceived to be due to evolutionary selective pressures for protective males. Secondarily, but similarly connected, these features are also considered to be related to the complicated sociosexual dominance interactions of terrestrial males. Generally the arboreal Old World monkeys show considerably less sexual dimorphism.

The hands of these monkeys are extremely dextrous. Almost all have **fully opposable** thumbs and thus quite good precision grips; this is especially true of the terrestrial species (Fig. 6–4). All digits have nails that are broader and flatter than those found in the New World primates.

Compared to New World primates, some Old World monkeys are "hardy souls" who can adjust fairly well to different climates, some as extreme as is found in Northern Japan. Their New World counterparts are comparatively fragile creatures who do not fare well in changing climatic and weather conditions. The considerable adaptive plasticity of the early catarrhines was probably an' important evolutionary feature for the present-day apes and man.

*Fig. 6–4* The excellent "pulp on pulp" opposability of thumb and index finger in a terrestrial Old World monkey, the baboon. This is essentially the precision grip. (S. I. Rosen.)

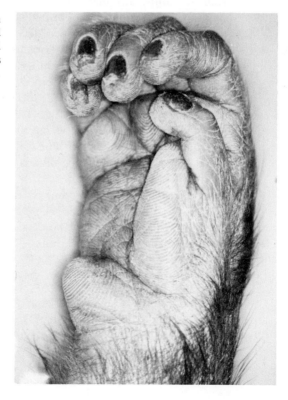

## The Colobines

The colobine (or leaf-eating) Old World monkeys are a rather primitive group of primates with species in both Africa and Asia. Of all primates, these are the most vegetarian. In part, their anatomies reflect their dietary specialization for leaves. While lacking the cheek pouches of the other Old World monkeys, the colobines have enlarged salivary glands in the mouth for more efficient predigestion of vegetation. More important, they have sacculated, or pouched, stomachs which provide an enormous surface area for contact and breakdown of the food. This type of stomach should not be confused with that of a ruminant such as a cow, which has essentially multiple stomachs. Some of the colobines are even capable of digesting dead fallen leaves from deciduous forests. These specialized stomachs, along with a very large intestinal tract, allow for maximum nutrition from relatively low nutritive vegetation. They do eat other vegetation, but leaves are their main course.

The colobines exhibit little to no sexual swelling and generally little sexual dimorphism. These traits are likely related to their basic arboreal habits, although some Asian species show some terrestrial behavior. Like all nonhuman primates, the colobines are basically quadrupeds; yet by locomotor specialization, they have been classified as brachiators, or more precisely, brachiating leapers. Their hands reflect such adaptation as do their limbs, the hindlimbs being much longer than the forelimbs. The colobines are more impressive leapers than brachiators, thus the hindlimb specialization rather than the typical brachiator forelimb lengthening. The splendid beauty of their leaping is rivaled in the primate world only by the brachiating aerobatics of the gibbon, an anthropoid ape. The colobines have very long tails which probably serve as balancing organs and airfoils in their leaps across vast areas between foliage.

The colobines exhibit an assortment of unusual nose shapes and sizes that is unparalleled elsewhere among the primates. The selective advantages, if any, of these noses are not evident. In some cases, they represent sexually dimorphic traits.

### THE COLOBUS MONKEYS OF AFRICA

The guerezas, or colobus monkeys, are the African members of the *Colobinae*. These brachiator-leapers are large monkeys who have slender bodies with unusually long and beautiful tails. In adaptation to brachiation, the colobus evolved a greatly reduced and sometimes absent thumb to achieve an efficient hook grip. Its thumb reduction is equal to that of the spider monkey in the New World. The colobus monkey likely occupies the same econiche in the Old World that the spider monkey does in the New World. The colobus is found throughout central sub-Saharan Africa. The colobus is a most strikingly beautiful animal (Fig. 6–5). White mantles adorn the sides of some colobus subspecies; these mantles continue into long, bushy tails. The pelts of these monkeys as well as their meat have made them the prey of man for centuries. The red colobus monkey is occasionally found in the diet of some chimpanzees.

Unfortunately we know very little of the behavior of the colobus monkeys of Africa. They appear to live in small groups of up to twenty animals, with small home ranges and moderate population densities. The black and white colobus seems to have two types of groups, one of mixed sexes and those which are all-male. Some of these monkeys use vocalization for spacing. There is definite use of branch breaking and quick movements to maintain social distance.

*Fig. 6–5* The king colobus monkeys (*Colobus guereza*). Note the completely white infant. (San Diego Zoo Photo.)

THE LANGURS OF ASIA

The Asiatic colobines are represented by a group of Old World monkeys collectively termed langurs. The common langurs (Fig. 6–6) are geographically distributed from India to China and Southeast Asia. These leaf-eaters are large arboreal creatures except for the Hanuman, or sacred langur of India, who spends a great deal of time on the ground. The Hanuman langur is also the largest of the langurs, possibly reflecting its terrestrial habits. The langur hand shows considerable adaptation to brachiation. The thumb is significantly reduced, while the fingers are very long.

Langur behavior has received a good deal of attention by a number of very competent investigators. The semiterrestrial Hanuman langur of India and Ceylon has received extensive investigation. Hanuman langur groups appear variable in composition, ranging from mixed sex groups, one-male groups (harems), to all-male groups. Harems and all-male groups tend to be found under crowded conditions.

*Fig. 6–6* The purple-faced langur (*Presbytis senex*). (San Diego Zoo Photo by Ron Garrison.)

Especially prominent among the North Indian langurs is the unusual permissiveness in the handling of infants. While males generally remain indifferent to the young, females freely pass their infants around allowing an unusual frequency of handling by other females—"aunt" behavior. It has been suggested by Dr. Phyllis J. Dolhinow that this permissiveness may be part of a socialization process that eventually allows for a variety of group compositions.

In recent years, primate ethologists have became more concerned with viewing primate behavior from not only the aspect of genetic (inherited) behavior but also behavior altered by ecological adaptation. Widespread studies of the Hanuman langur have revealed interesting evidence. Where the environment is harsh, harems and also stray males are common. Male dominance is not well defined and rather unstable. Also the more severe the environment, the more aggressive the langur behavior. There also seems to be a direct relationship between the size of the home range of the group and aggressive behavior—the larger the range, the more peaceful the relations between the Hanuman langur groups. In Ceylon where the home ranges are quite small and crowding occurs, the behavior is very aggressive and fights are quite common. The semiterrestrial Hanuman langurs, unlike the savannah baboons, have no dependence on large aggressive males for protection. The langur aggressive behavior is not the result of a dominance hierarchy in the group but is a behavioral adaptation to severe environments and crowding.

Another type of langur and the most unusual looking of all Old World primates is the proboscis monkey of Borneo (Fig. 6–3). The nose of the male proboscis almost defies adequate description. Bulbous, pendulous, this nose hangs over the mouth and seems to obscure the face. Appropriately, this animal is placed in the genus *Nasalis*. The significance of this type of nose may be sexual, since females and infants have noses that are less well developed and more bobbed or upturned. It probably plays no major role in the sense of smell, because the proboscis monkey has no better olfactory development than other langurs. More likely it is a resonation chamber for the "honks" of the male. The male proboscis is also much larger and heavier than the female; this is unusual for colobines. These red-colored langurs often come to the ground. This partial terrestrialism may play a role in their sexual dimorphism. Proboscis monkeys are good swimmers, preferring at times to swim considerable distances underwater. Like all colobines, these langurs sit and sleep in erect postures.

The remaining Asiatic colobines are sometimes collectively called the snub-nosed langurs. The snub-nosed langur of China and North Vietnam has become acclimated to cold weather. The noses of these creatures appear at times to be virtually absent. Another type of snub-nose is the Pagi Island, or pig-tailed langur, which is reported to inhabit some islands off the western shores of Sumatra. It is so rare a primate that few people have ever seen it. It is reputed to have a short tail, which is very unusual for a colobine.

## The Cercopithecinae

Some fifteen million years ago there may have been a rapid adaptive radiation from a generalized Old World monkey type. Populations of this type evolved into a number of arboreal and terrestrial econiches in Africa and Asia now occupied by the *Cercopithecineae*. For simplicity, we divide the Cercopithecineae into two groups: the Asian macaques and the African baboons, guenons, and mangabeys. These are the so-called cheek-pouched monkeys. These expansions of the oral cavity allow this type of Old World monkey to shove and store large amounts of food into its mouth while at the same time being able to eat. If the monkey has to flee a grazing or browsing site, he is assured the retention of some food in these pouches. He is similarly assured food in case he has to give way to a more dominant member of his group (see Fig. 6–16).

### THE GUENONS

We shall consider the African guenons first because they are probably closer to the ancestral population of the *Cercopithecinae*. The Guenons have an extremely wide distribution throughout Africa. These creatures have evolved into countless species and subspecies (Fig. 6–7 and Fig. 6–8). Most of them are wholly arboreal. Although almost every kind of monkey is capable of some brachiation, the guenons are best described as arboreal quadrupeds. Their hind and forelimbs are of approximately equal length. These monkeys do not tend to be large; they come in a variety of muted colors, often with striking facial hair in the form of brightly colored eyebrows, moustaches, and beards. The male external genitalia is as a rule very bright in color, yet there is almost no sexual swelling in either males or females.

The guenons have been able to adapt to a number of different environments—rain forest, forest swamp, and semiarid land. In certain parts of Africa, especially West Africa, it is not unusual to find mixed-species groups of guenons, mangabeys, and colobus monkeys. These different species seem to be able to live among one another without considerable tension. This may be due to each species' particular dietary

*Fig. 6–7* De Brazza's guenons (*Cercopithecus neglectus*). (San Diego Zoo Photo by R. Van Nostrand.)

preferences. They may also derive mutual protection from each other's defense behavior. Guenon groups vary from one-male groups (harems) to groups numbering up to fifty; but generally they are found in small groups. In spite of small groups, they maintain moderately large population densities.

*Fig. 6–8* The moustached guenon (*Cercopithecus cephus*). (San Diego Zoo Photo by Ron Garrison.)

One type of guenon, the vervet, is primarily terrestrial and is reputed to show a proclivity for erect postures. The behavior of the vervet has begun to receive considerable attention in recent years because of its behavorial plasticity. Vervets occur in varied environments—the more severe the environment, the more apparent the sexual dimorphism. Multimale groups are found in the harsh environments, but there is also a trend for more females in this setting. Territoriality also seems to be more extreme in the harsher environments. Vervet behavior reveals that guenons, at least the vervets, have an inherited behavioral capacity for adapting to a number of environmental settings.

Another ground dweller is the Hussar, or patas monkey (Fig. 6–9). Although it has been said that the patas monkey is a form intermediate between the arboreal guenons and the ground-living savannah baboons, it is closely related to the guenons. The relationship is so close that some primatologists place this creature in the genus *Cercopithecus*. The patas monkey is a large monkey of reddish color who lives in high grass country. The patas often resorts to erect postures and bipedalism, in order to look over the tall grass for predators. He can then act as a decoy while the females and subadults freeze in the background and subsequently run off. The male patas also employs a stereotyped jumping on "all fours" behavior which probably serves as a threat gesture and also a decoy mechanism. Males will also use arboreal display behavior for decoy purposes. The patas monkeys exhibit noticeable sexual dimorphism, while the primarily arboreal guenons do not. A large colorful male makes an excellent decoy. The patas body conformation has been at times called "doglike" because the limbs are quite long and roughly equal in length; the male's behavior has been compared to that of a watch dog. The patas is the fastest primate on the ground, attaining running speeds up to thirty-five miles an hour. Obviously this running ability gives the patas an additional advantage in defense behavior.

*Fig. 6–9* The patas monkey (*Erythrocebus patas*). Note the military moustache. (San Diego Zoo Photo.)

The patas monkey is one of three Old World monkeys (hamadryas baboon, and gelada baboon) that inhabit arid open-country areas. A good deal of their social organization and behavior is geared to survival in such an environment. The patas monkeys live in small groups with very low population densities. Their home ranges are very large, thus allowing maximum exploitation for their omnivorous diet. As in the case of the gelada and hamadryas baboons, the social unit is the one-male group. Males without harems are forced to form all-male groups. Patas males with harems will not tolerate other adult males in the group. A good deal of male patas behavior is centered around maintaining distance from other males and harems. This is not territorial behavior in the conventional sense. Even if the environment could support greater numbers of patas monkeys, the male behavior would probably be the same, since it is genetically derived.

*Fig. 6–10* The sooty mangabey (*Cercocebus atys*). Note the eyebrow flashes. (San Diego Zoo Photo.)

## THE MANGABEYS

The mangabeys are large, slender, long-tailed monkeys who have decorative faces with white eyebrow flashes and prominent cheek tufts (Fig. 6–10). They are divided into two groups: one a completely arboreal type and the other primarily a ground dweller. Both types live in the African tropical rain forests, one foraging on forest floors, the other dwelling in very tall trees. Sexual dimorphism is minor in both types.

## THE BABOONS

The baboons represent the extreme ground-dwelling nonhuman primate. Except when resting, sleeping, or fleeing danger, baboons are found on the ground. These primates meet the standard concept of the extremely sexually dimorphic terrestrial primate. The adult male baboon is grossly larger than his female counterpart, sometimes twice as large. He has enormous canine teeth and large prognathous (forwardly projecting) jaws for defense (Fig. 6–11). The face and jaws form a massive muzzle, atypical of a higher primate. The adult male is also formidable in appearance due to a profuse mantle of fur on his head and upper body, much like a lion's mane. Being so terrestrial, baboons, including the drill baboon but not the mandrill, are of dull color. (The colorful primates tend to be arboreal.) The ischial callosities of these creatures are proportionately much larger than those of other Old World monkeys. In addition, an unusually large area of soft-tissue padding surrounds their callosities. The female has an extensive sexual skin area that becomes very inflamed at oestrus.

The diet of the baboon is mostly vegetarian—grasses, their roots, and other tubers. Seasonal fruit is also eaten on occasion, as is meat in the form of small animals such as hares, infant gazelles, and lizards. Even with their magnificent canine teeth (which are primarily for "show," although fighting and wounding does occur), it would be impractical for baboons to be carnivores, since only a few large males could be effective hunters and probably only for themselves. Their absence would also leave the rest of the baboon troop very vulnerable to attack by predators.

The baboons are geographically remarkable, for they occupy almost a continent; they are found throughout sub-Saharan Africa and possibly a small part of the Middle East. They live in semideserts, tropical rain forests, savannah grasslands and woodlands, and even on rocky sea cliffs. Again we have dramatic testimony to the adaptability of Old World species. The common, or savannah baboons (olive, yellow, Guinea, and

*Fig. 6–11* An adult hamadryas baboon (*Papio hamadryas*) in threat gesture. Note the hairy mantle. (San Diego Zoo Photo.)

Chacma) are probably not distinct types but examples of intergrading geographical races; clines definitely exist for some baboon traits.

For almost a decade primatologists spoke of "baboon behavior." This behavioral information came largely from the now classic work on the savannah baboon by Irven DeVore and S. L. Washburn. Today we speak of "ranges" of baboon behavior. We now know that the type of behavior described by DeVore and Washburn holds true primarily for savannah baboons but not all baboons, notably the hamadryas and gelada baboons.

The savannah baboon displays a behavior that has evolved from living in varying degrees on and near savannah grasslands but always near trees, which afford protection and rest. The social structure of the savannah baboon is the troop; this is the center of its universe. The troop is organized in an extremely orderly fashion. By the time it is an adult every savannah baboon will know its role and rank in the troop.

Of prime importance to the savannah troop is the male central hierarchy. This hierarchy is composed of several males of which one is the most dominant. This dominant male is often the largest, or at least the one with the largest canine teeth. He is dominant over the other members of the central hierarchy and the troop as a whole, although two or more subordinate males in the hierarchy together may be able to displace him at times. Most dominance interaction is by gesture, especially threat gesture and mock combat. Occasionally injury does occur, but the gestures are so ritualized that this is not the rule. This central hierarchy system is extremely important to the survival of the troop, which may range to over one-hundred baboons but usually averages around forty animals. The male central hierarchy protects the troop against predators such as lions. One dominant male with large canines is not as formidable as several large males. This central hierarchy, especially the dominant male, also directs the troop move-

*Fig. 6–12* The male mandrill baboon (*Papio-Mandrillus-sphinx*). (Courtesy of Ploydor, Inc.)

ments. The central hierarachy and the dominance status system also play important roles in the reproductive continuity of the troop. The most dominant male has exclusive priority to an oestrus female, if he desires to form a consort pair with her for mating. As long as he remains with her, he alone has sexual access to her. After the consort pair disbands and also before it is formed, subordinate males have access to the female. The important point here is that the dominant male genes are kept flowing in the troop. This does not mean that all newborns are the offspring of the large dominant male and the most dominant female in the troop. The system does insure that some offspring will be his and that the troop will always have some large males. The troop assures each baboon leadership, protection, knowledge of food and water locations, and mating.

### THE MANDRILL

The mandrill baboon is an inhabitant of equatorial West Africa. Evolutionarily these populations may represent baboons which have made a secondary adaptation to rain forests and mountainous regions. The male mandrill is the most colorful of all primates (Fig. 6–12). His massive muzzle has long multiple grooves and "swellings" which run from below the eyes to the edge of the nose; this area is of a lilac color, while a strip of bright red flesh-colored skin runs down the middle of the muzzle and expands onto the nose. This same basic color pattern is seen on his external genitalia. In contrast, the female, who also has these swellings, is a dull-colored, small creature who lacks even sexual coloration (Fig. 6–13).

### THE DRILL

The drill baboon is a duller colored West African cousin of the mandrill. He sports a Lincoln beard, which gives his head a formidable appearance (Fig. 6–14). While not sharing the elaborate facial coloration of the male mandrill, the male drill has genitals that are even brighter than those of the mandrill.

### THE HAMADRYAS BABOON

The hamadryas baboon is the sacred monkey of ancient Egypt; today it is found mainly in Ethiopia. This animal has a flesh-colored face and a high bobbed nose. (Fig. 6–11). Very notable is its long fur, especially in the male. The hamadryas baboon is adapted to desertlike grassland and takes refuge and sleeps on rocky cliffs. On a comparative basis, these baboons are distinct socially, and to some extent biologically, from the common baboons.

*Fig. 6–13*  The rather colorless female mandrill baboon and offspring. (San Diego Zoo Photo by Ron Garrison.)

*Fig. 6–14* A female drill baboon (*Papio leucophaseus*) and infant (Courtesy of the Zoological Society of Philadelphia.)

To a large extent our concept of the range of baboon behavior has been changed by the information gathered on the behavior of the hamadryas baboon. The excellent studies of Dr. Hans Kummer have brought this new information forward.

Instead of the troop being the main social unit, as in the case of the savannah baboon, the harem is the major social unit of the hamadryas baboon. Approximately 20 percent of the adult male hamadryas baboons are "bachelors." The male is very covetous of his females; he will bite the backs of their necks to "keep them in line." This herding behavior is not seen in the patas monkey and gelada baboon harems. Hamadryas males do not take females from other harems. Kummer has pointed out what might be termed "pseudofemale" behavior by male harem leaders. They will carry young infants possibly to transfer maternal status to themselves so that subadult females will be attracted to them. In addition, Kummer has raised the possibility that the extremely furry mane of the hamadryas male serves as a grooming device to attract young females.

The males of harems will at times unite into a kind of superband, a second type of hamadryas social unit. These superbands will challenge other bands. A third type of hamadryas social unit is the troop, which, of course, is made up of the harems and bachelor males. The rocky sleeping cliffs harbor a fourth type of social unit, a sort of supertroop. In areas where cliffs are scarce, such groupings can number up to 700 baboons. Thus the hamadryas baboon has a fusion-fission type of social structure. This is obviously well suited to an environment where food and water sources, along with roosting places, may be in short supply.

## THE GELADA

The gelada baboon is a most interesting and unique Old World monkey (Fig. 6–15). In fact, some primatologists do not consider the gelada a baboon at all but a type of macaque Old World monkey. It is probably best considered a macaquelike baboon. The macaques and baboons are really not separate biological entities but variations of one type of Old World subpattern; the two groups grade into one another.

The geladas are found in the grassy and mountainous rock areas of Ethiopia; they are even reputed to be in the southern part of Arabia. They represent a unique adaptation to that type of environment. They are the most terrestrial of all baboons. Dr. C. J. Jolly has offered a useful classification of this type of adaptation. Jolly considers the gelada to be what he calls a **small-object feeder**. While the other baboons eat large grasses and roots, the gelada consumes smaller vegetation such as seeds. Jolly terms this diet **graminivorous**. The results of adapting to this have been termed the **T-complex** ("T" for *Theropithecus*). The T-complex is a particular type of anatomy that features shorter and smaller jaws with modified molar teeth. Jolly believes early hominids (man) may have been small-object feeders.

Geladas are large monkeys with hour-glass-shaped faces. The upper sides of their muzzles have grooves similar to the mandrill. The nose is extremely bobbed. Quite unique is the possession by both males and females of sexual chest patches. The male's is somewhat callused and

*Fig. 6–15* A male gelada baboon (*Theropithecus gelada*) grooming a female. Note the sexual skin patch of the female and the juxtaposed nipples. (San Diego Zoo Photo by Ron Garrison.)

triangular, while the female has one shaped like an hour-glass and surrounded by round bumps (caruncles) that are sometimes called a "pearl necklace" or "rosary beads." The female's breasts are pendulous and meet in the midline of the chest. At oestrus, her patch becomes bright red, sometimes even more brilliant than her rump patch sexual swelling. Dr. Wolfgang Wilder has suggested that the external chest anatomy of the female gelada is an evolutionarily selected replica of the genitalia. The female's rump patch also has caruncles. The chest patch serves as a visual sexual cue. A similar but lesser case can be made for the male.

The behavior of the gelada has received more attention in recent years because of the gross differences of its behavior from that of the savannah baboon and the apparent concergence with aspects of hamadryas baboon behavior. The geladas live in open, arid country; this environment can be severe at times. In adaptation to such an environment, a particular type of social organization evolved, the fusion-fission type of primate society, which is also seen among the hamadryas. In times of plenty, usually the wet season, harems and all-male groups will fuse into large troops numbering up to 400 members. When there is a short supply of food in the dry season, smaller troops are formed. If conditions are harsh enough, there may not be a recognizable troop at all.

Unlike hamadryas males, gelada males do not herd females. To some extent the dominance role of the male is taken over by the most dominant female in the harem. She herds the females, to some extent, maintaining them in the harem. The gelada male, who has rather small canine teeth, is protective of his females, but he defends only his sexual territory and no other. He also seems to play a protective role when the group moves to its rocky cliffs for sleep or when fleeing danger. His role here is to block the path of access from below.

THE MACAQUES

The macaque Old World monkeys are comprised of more than a dozen different species representing part of a broad spectrum of Old World monkeys. Only man has a greater geographic distribution than the macaques. They range from Afghanistan to Southeast Asia and its islands, and to China, and Northern Japan. Primarily Asiatic in habitat, a tailless member of the macaque family called the Barbary ape is found in North Africa and on the Rock of Gibraltar, where it was introduced by man. Legend has it that as long as the Barbary ape remains at Gibraltar so will the British. One of the duties of the British regiment at Gibraltar is to insure the survival of these monkeys. Ironically, the Barbary ape once roamed as far as England in prehistoric times. For fourteen centuries, human anatomy was based upon the Greek physician Galen's knowledge of the anatomy of this macaque. Today, the Barbary apes are a tourist attraction at the Rock.

The macaques represent a kind of intermediate Old World locomotor form. Many are both arboreal and terrestrial. It is possible that man, the agriculturalist, in clearing land has forced the macaque (once wholly arboreal) into partial ground-dwelling niches. A good example is the rhesus monkey of India, who inhabitis villages and temples; he is also often seen as a street begger. The term "urban monkey" has become synonymous with the rhesus. This macaque is very familiar because of the large role it has played in biomedical research and for the discovery of the rhesus blood factor.

The macaques in general are rather large monkeys of dull color. Especially large in size are a group sometimes called the stump-tailed macaques; the Barbary ape is sometimes included in this group. In this group are: the stump-tailed macaque of mainland Southeast Asia, the pig-tailed macaque of islands of Southeast Asia, and the so-called Celebes black ape (Fig. 6–17) and Moor macaque (Fig. 6–16) of the Celebes. Dr. Jack Fooden has shown that the Moor macaque and Celebes black ape are actually two extremes of a spectrum of macaque variation on the Celebes land mass. The Celebes black ape is a baboon-like macaque that represents a morphological crossing point in the macaque-baboon range of Old World primate variation.

Quite interesting among the stump-tailed macaques are the Japanese macaques, who have been extensively studied by the Japan Monkey Center. Like most Old World monkeys, the Japanese macaque is a very adaptable animal. In the northerly reaches of Japan, on the Shimokita Peninsula of Honshu Island, live the Japanese macaques, who have become known as snow monkeys. These macaques have become adapted to the extremely harsh winters of northern Japan. Like all Japanese macaques, their fur is quite thick; in some cases lighter fur colors have likely been selected for. In winter, these animals live off tree bark, a food heretofore only consumed by some Lemuriformes. Some of these macaques temporarily take relief from the great cold by bathing in natural hot springs.

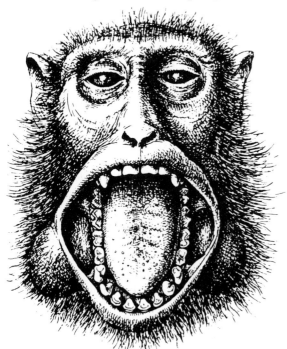

*Fig. 6–16* A demonstration of the cheek pouches of a *Cercopithecinae* monkey, the Moor macaque (*Macaca maurus*). (Courtesy of A. H. Schultz.)

*Fig. 6–17* The Celebes black ape (*Macaca niger*). San Diego Zoo Photo.)

Perhaps the most spectacular findings on macaque behavior have been made by primatologists studying the Japanese macaques of Koshima Island, Japan. This isolated spot has been an ideal locale for observing and experimenting with macaque behavior. Several years ago sweet potatoes were placed on the Koshima beach to serve as rations for the macaque troop there. A number of unexpected events took place. A young female macaque of approximately two years not only took to eating the

*Fig. 6–18*    A lion-tailed macaque, or wanderoo (*Macaca silenus*) in a threatening gesture. (San Diego Zoo Photo by Ron Garrison.)

potatoes but secondarily invented washing the sand off them in a stream. Later she shifted to potato washing in the sea. The new tradition of potato washing spread from this young female to other youngsters and eventually to adult females. The adult male macaques took no active role in the tradition, giving new insight into the adult male macaque behavior—which to some extent appears to be more programmed than that of females and their young. However, additional tradition arose. The potato-washing macaques acquired a preference for sea-salted potatoes. More important, previously water-shy adults became sea waders. This water behavior was later passed on to newborns to the point where it has become a normal part of the infant socialization process.

Some time after the adoption of potato washing, the Koshima investigators placed wheat on the beach. Again the same female who invented the potato washing tradition began a new tradition—the water cleaning of wheat. She placed the wheat on the water; while the wheat floated, the sand sunk to the bottom. Again a behavioral tradition was invented. Because of the inventive nature of this behavior and its becoming a learned tradition, it has been termed "preculture," a rather undefinable concept. Such behavior is to a certain extent seen in some baboon food-cleaning habits. While this all may give us some possible insight into how early human behavior evolved, it more clearly points out that much of nonhuman primate behavior is so adaptable that, given the correct circumstances, new behavior can be invented and transmitted. The Koshima experience also indicates that behavioral experiments can be created to help us separate purely genetic behavior from singularly adaptive behavior.

The macaque varieties are too numerous for all to be considered in this brief description. Generally their wide variety is thought to be a relatively late adaptive radiation. They illustrate well that the primate pattern is so generalized that all types of econiches can be adapted to and survived in by primates who have retained most of the primate pattern.

## SUMMARY

CATARRHINES
*Cercopithecoidea*
*Colobinae*
LEAF-EATERS
*Cercopithecinae*
CHEEK-POUCHED MONKEYS
FOREST-DWELLING BROWSERS
GRASS-EATING SAVANNAH DWELLERS
THE OLD WORLD MONKEY PATTERN
    Wide Distribution
    Comparative Uniformity of Anatomy
    Catarrhine Noses
    Dental Formula $\frac{2.\ 1.\ 2.\ 3.}{2.\ 1.\ 2.\ 3.}$
    No Prehensile Tails
    Diurnal
    Robust
    Ischial Callosities
    Variety of Sexual Swellings
    Single Births
    Biologically Closer to Man
    Sexual Dimorphism Quite Common
    Fully Opposable Thumbs

SMALL-OBJECT FEEDER
GRAMINIVOROUS
T-COMPLEX

Chapter 7

# The Anthropoid Apes

The "highest" of the higher primates are the ***Hominoidea***. This super-family is made up of three separate families: (1) the ***Hylobatidae*** or hylobatids, more commonly known as the **gibbons** or by the misnomer "lesser apes"; (2) the ***Pongidae***—pongids, or **great apes**; and (3) the ***Hominidae*** or hominids—that is, **man**. These are, of course, convenient categories imposed by the mind of man; they are not necessarily natural categories and are by no means definitive.

What then, is an ape? If we define an ape as a "tailless anthropoid," we must then place man as an ape. Naturally, this is offensive to our human nature, even though man and ape are separated more by ego than anatomy. Granted there is an ape grade of primate organization, but it is much less distinct a grade from man than the monkey grade is from the prosimian.

## The Hylobatids

Most authorities recognize the hylobatids as being a group of apes distinct from the pongids. We know that the hylobatids have had a very long and separate evolutionary history from the great apes. The hy-

*Fig. 7–1*   One of the Common gibbons (*Hylobates agilis?*). (Courtesy of Animal Talent Scouts, Inc. Photo by Ralph Morse, *Life* magazine, © Time, Inc.)

lobatids are commonly divided into two types: the **common gibbon** (Fig. 7–1) and the **siamang gibbon** (Fig. 7–2). The common gibbons are made up of half a dozen species and unnumbered subspecies; it is almost an impossible task to tell one common gibbon from another, because individual variation is so great.

The gibbons are found in the forests of Southeast Asia including Sumatra, Borneo, and other islands of Indonesia. The siamang gibbon is a distinct animal which some authorities believe is closer to the great apes than the common gibbons. Found in Sumatra and the Malayan

*Fig. 7–2* The siamang gibbon (*Symphalangus syndactylus*). Note the grasping feet and the size of the vocal sac. (Courtesy of the Zoological Society of Philadelphia.)

Peninsula, the siamang is of heavier build, shorter stature, and possesses a less dense but shaggier coat than the common gibbons. The siamang is further distinguished by its invariable black color, large, external laryngeal air sac, and webbing between the second and third toes.

## The Hylobatid Pattern

Of all the primates, the gibbon is the most fully adapted to brachiation. Except for 10 to 15 percent of locomotor time, the gibbon is a **full-time brachiator**. The gibbon is sometimes called the "true brachiator." This **tailless** ape is the model by which other primates are judged anatomically and functionally as brachiators. The skill and beauty of the gibbon's brachiation cannot adequately be described, even still pictures do not do it justice (Fig. 7–3). Its grace and agility rival that of gazelles and gliding birds. Almost its entire anatomy is designed for brachiation (Fig. 7–4). The **forelimbs** are **extremely long**, the arm length being

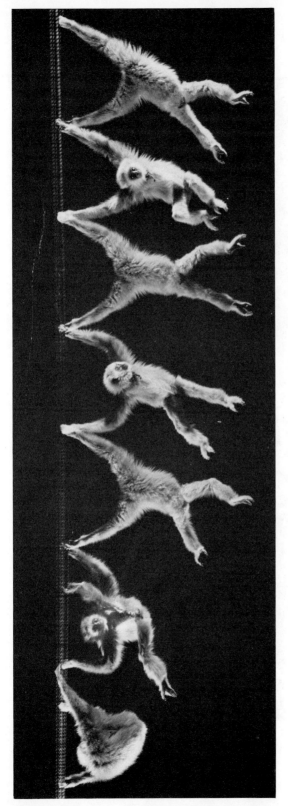

*Fig. 7–3*  Time-lapse photo of a brachiating gibbon. Note the hindlimbs are not extensively involved in the locomotor pattern. (Courtesy of Animal Talent Scouts, Inc. Photo by Ralph Morse, *Life* magazine, © Time, Inc.)

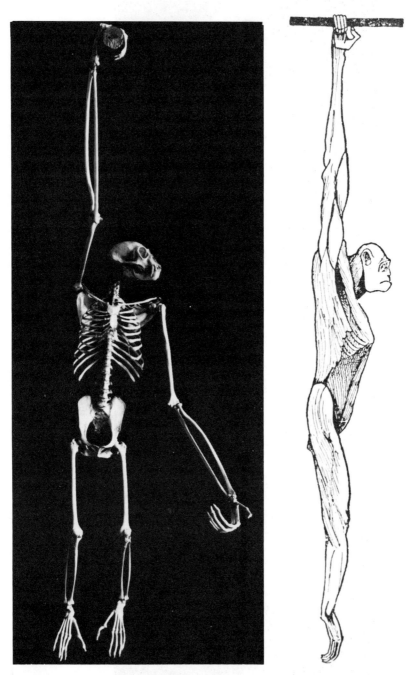

*Fig. 7–4*  The full-time brachiator's anatomy. (Left—courtesy of the American Museum of Natural History; right—after Keith.)

approximately two and one-half times its trunk length. Contrary to popular belief, the gibbon actually has long hindlimbs, which cannot be fully extended at the knee joint (Fig. 7–19). The hindlimbs are tucked up to the trunk in brachiation. These factors, along with the forelimb length, give the hindlimbs a deceptively short appearance. The gibbon's hand is very long and narrow; surprisingly, the thumb is quite long. The thumb is placed far back on the wrist and is turned across the palm to effect a hook grip in brachiation. Unlike the great apes, the hylobatids have a good precision grip.

Ecologically and anatomically, the gibbons occupy the econiche held by the spider monkey in the New World; behaviorally this primate may also occupy a niche similar to that of the howler monkey, especially in light of the loud hoots the gibbon voices in the morning. The siamang not only hoots but also carries on between hoots an alternating booming sound in his vocal sac. This sac can be inflated to sizes larger than his head (Fig. 7–2).

These most specialized of brachiators eat leaves, young birds, and insects but are mostly **frugivorous** (fruit-eaters). In the middle canopy of the forests, they engage in what we call terminal-branch feeding; their brachiating may represent an evolutionary adaptation to gathering fruit on the ends of branches. When not brachiating, gibbons climb and walk quadrupedally in the trees. They also walk erect bipedally atop tree limbs; in fact, the genus name *Hylobates* means "tree-walker." If placed upon the ground, the gibbon will generally prefer semierect bipedal locomotion. When walking bipedally, the gibbon's arms are raised outwardly above the head with the hands bent at the wrist, the upper limbs acting as balancing organs. When standing semierect his hands will flop on the ground due to the extreme forelimb length. Some primatologists see many of the gibbon's brachiation-related adaptations as preadaptations for erect bipedalism. The gibbon's trunk is erect in brachiation. Like all primate brachiators, the gibbon has a very wide, yet shallow, thorax (chest cavity) and a forwardly placed spinal column. These factors set up weight-balance-gravity relationships that are necessary for sustained erect posture. It should be noted that in its natural environment the gibbon is **wholly arboreal.**

For apes, the gibbons are small—actually **monkey sized.** They present little body mass relative to their maximum three-foot height. A heavy gibbon is one over twenty pounds. Also unusual is the fact that they are the only apes who have **ischial callosities always present.** It is now known that there is a cline for ischial callosities among gibbons. The callosities become smaller as one travels northward in their habitat.

These most numerous of all anthropoid apes exhibit **little sexual dimorphism.** Body sizes and weights of males and females are remarkably similar. Both males and females have very long, projecting canines, the male's being slightly larger. These teeth may be adaptations for piercing the outer shell of fruit. Although there are no sexual swelling evident in either sex, the female gibbon does appear to have an oestrus cycle and is sexually receptive only when in "season." It has been shown by Dr. Jack Fooden that coat-color differences exist by sex (sexual dichromatism) in three species of the common gibbon.

The English anatomist and paleontologist, Sir Richard Owen (1804–1892), coined the term "brachiator" to describe the gibbon. Later, the English anatomist and physical anthropologist, Sir Arthur Keith, popularized the terminology of brachiation. Keith considered the gibbon to be a primitive Old World monkey whose anatomy had secondarily adapted to (orthograde) erect postures. As we will see in Chapter Nine, Keith may have not been too far from the truth concerning the fossil ancestors of the modern gibbons. A few modern primatologists favor placing the gibbon in with the Old World monkeys.

As mentioned earlier, the gibbon is one of the few primates that lives in a monogamous state. We owe most of our knowledge of gibbon behavior to the studies Dr. C. Ray Carpenter made in the 1930s and to the more recent work of Dr. John O. Ellefson. The gibbon nuclear family is made up of a male, a female, and any offspring, which may number up to four. Thus a large gibbon group would have six members. Subadults of either sex are driven from the group by the corresponding parent—i.e., the father drives out the maturing son. This ostracism is the foundation for new families.

Gibbon families live in quite small areas but have relatively low population densities. The territory of a gibbon family is essentially its home range. The group virtually always retains the same territory and will sleep near the center of that territory. Gibbon territories do overlap, and that is where territorial behavior is at an optimum. While it appears that the howler monkeys are able to keep other howler groups away by avoidance calls alone, the gibbon's territorial behavior involves true aggression and visual contact. Gibbon groups know precisely where their territories overlap. The male gibbon's "hoo" is an attraction call, and also provokes visual attention from other gibbons. Ellefson has recorded true aggression by gibbon males upon one another—mainly grabbing and biting. However, compared to that of baboons and macaques, gibbon aggressive behavior is at a very low level. The female gives what has been called the "great call," which probably serves as a location call to other groups and also her own family. Gibbon behavior is programmed for the protection and retention of a rather precise arboreal territory. The maintenance of this territory insures an adequate food supply for the family. The infant gibbon grows up without a peer play group; his or her behavior is not only genetically determined but re-enforced by parental example. If the gibbons have had as long an evolutionary history separate from the other apes as we believe, such precise social organization and behavior is understandable.

### The Pongids

The great apes, or pongids, are a hominoid group composed of three distinct primates: the **orang-utan** of Southeast Asia and the **chimpanzee** and **gorilla**, both of Africa. The orang-utan is confined to the large islands of Sumatra and Borneo. The chimpanzee is a native of equatorial West and Central Africa and part of East Africa as far as Lake Victoria. Chimpanzees are found throughout the area. Also native to equatorial Africa, the gorilla is now confined to two separate areas of Central Africa, one in the western part and the other in the east. While the orang-utan is not generally considered to be closely related to the chimpanzee and gorilla, some primate biologists consider these latter two animals to be genetically similar.

### The Pongid Pattern

While the total morphological pattern presented for the living primates covered to this point has been presented rather concisely, the pongid pattern will be presented more extensively, since most primatologists consider these primates to be man's closest living relatives. Another popular belief is that the great apes and man are derived from a common ancestor from the Miocene. Most often and perhaps too freely, our fossil ancestors are compared to the pongids and to modern man to determine our ancestors' "humanness."

The pongids are the **largest** and **heaviest nonhuman primates**. In comparison to the other nonhuman primates, one can safely term these creatures "giant primates," as is man. Their dietary requirements are considerable, even though large mammals tend to have relatively lower basal metabolic rates than smaller ones. Their diets are bulky, since they are primarily **vegetarians** and ingest little or no animal protein. Except for the gorilla, they are **largely frugivorous**. Their dentition greatly reflects their diet. **Large projecting** conical **canines** are used to open fruit and strip vegetation. Like most primates with projecting canines, a small gap, or **diastema** is present in front of the upper canines and behind the lower canines so that the projecting canine tooth from the opposite jaw will have a space in which to fit (Fig. 7–5).

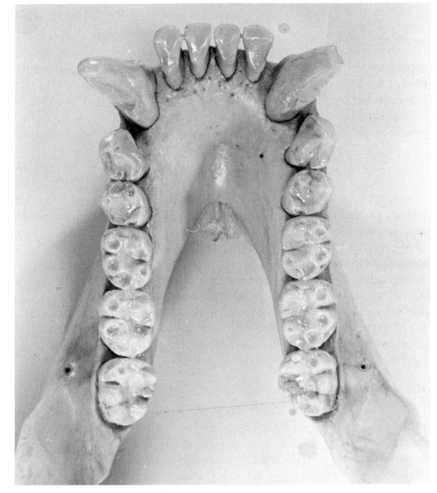

Fig. 7–5  A gorilla lower jaw. Note the diastemata, simian shelf (*SS*), and the U-shaped dental arcade. (S. l. Rosen.)

Fig. 7–6  The molar tooth of a hominid, man (left) and an Old World monkey, rhesus (right). The monkey molar anterior cusps (and the two posterior) are joined by a transverse ridge. The separate cusps of the human molar are separated by Y-shaped valleys—the Dryo Y-5 pattern. (J. Shea.)

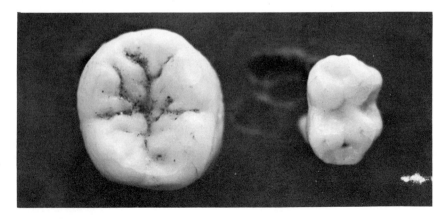

The incisor teeth are slightly procumbent—an asset in opening the outer shell of fruit. The first premolar tooth is what we call "sectorial," that is, it has cutting edge. Their molar teeth, like man's and the gibbon's, have **separate cusps**. The Old World monkey molar teeth have cusps united by transverse ridges (Fig. 7–6). The pongid molar cusps are also high and rather sharp, a vegetarian adaptation. Man's molar teeth have short, blunt cusps. The **jaws** containing this type of dentition are **massive** and **prognathous** (forwardly projecting). The teeth are arranged in these jaws into a **U-shaped dental arcade**. Since the pongid has **no chin** to buttress the mandible (lower jaw) against the extreme pressures applied to it by the massive muscles of mastication and the bulky foods consumed, a substitute buttress is present. This is the **simian shelf** (Fig. 7–5). This shelf of bone is seen also in some monkeys. The large heavy dental apparatus is further reflected in the **extensive cheek bones** (zygomatic region) of the pongids. One of the chief muscles of mastication (the masseter) originates here, and the region also acts as a buttress, absorbing the great pressures from mastication.

Another chief muscle of mastication (the temporal muscle) needs an extensive area of origin in the pongid. In the case of the large-jawed males, the temporal muscles are often so massive, and the braincase so small, that a **sagittal crest** on top of the skull is required to further anchor the temporal muscles at their origins. The sagittal crest is usually restricted to males—a sexually dimorphic trait. This structure is directly related to the size and weight of the jaws and the size of the brain case, which are for the most part genetically determined. Thus, the presence of a sagittal crest in the male gorilla, male orang-utan, and occasional male chimpanzee is to a certain degree a genetic trait. Some monkeys also have them.

The pongids generally have **massively developed supraorbital regions** (browridge area). In some cases this region forms a continuous (uninterrupted) bar of bone, a supraorbital torus. In other cases, there are separate supraorbital ridges (Fig. 7–7). These configurations are also found in some monkeys. Structurally these well-developed supraorbital regions are related to the fact that the pongids **lack a forehead**; the frontal area of the brain is meagerly developed. Functionally this region's development is partially related to the stresses from mastication that are applied to it and the occasional attachment of the temporal muscle fibers.

Considerably more than any other nonhuman primates, the pongids can maintain **prolonged semi-erect postures** and **bipedal locomotion**. This "anatomical behavior" is prolonged compared to other primates but not man. Like the gibbons, pongids (except for a few rare individuals), cannot fully extend their knee and hip joints; thus fully

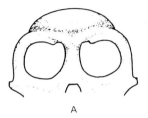

A                                          B

*Fig. 7-7*   The two major types of robust primate supraorbital conformations: (*A*) interrupted supraorbital ridges and (*B*) the confluent supraorbital torus. (After Schwalbe.)

erect posture is not possible. In order to support their large, heavy skulls, the pongids have extremely well-developed nuchal (neck) muscles. This musculature is attached on the back of the skull to **large occipital crests.** In spite of proficiencies at semierect postures, the pongid has a backwardly placed foramen magnum and occipital condyles on the skull base. The pongid also possesses the basic **quadrupedal pelvis,** which is slightly expanded in the gorilla. Like the gibbons, the pongids are **tailless.** The musculature that moves and supports the monkey's tail is rearranged and enlarged in the apes and man to support the pelvic viscera in upright posture.

Absolutely and relative to body size, the pongids have **very large** and **complicated brains.** The largest cranial capacity documented for a nonhuman primate to date is 752 cc (cubic centimeters); this was calculated from a large gorilla skull by Schultz. These pongid brains are not as large, absolutely and relatively, and as complex as man's brain; however, they do approach the neuroanatomy of man more closely than other primates.

While all of the pongids do not regularly brachiate, they are all **structural brachiators.** They possess **relatively long forelimbs** and **reduced thumbs.** The thumb is usually so reduced in length that the pongids are only of **poor precision grips.**

Similar to man, the great apes have significantly **prolonged life spans.** Evidence indicates lifetimes of over forty years are not uncommon. Again like man, **delayed maturity** is exhibited. The first two years of a pongid's life are behaviorally and developmentally quite similar to that of the human infant. When full maturity is reached, **considerable sexual dimorphism** is often evident in body size, weight, and conformation.

Unlike the hylobatids, the great apes' possession of **ischial callosities** is **infrequent.** Like the gibbons, but unlike Old World monkeys, the pongid ischial callosities, when present, are developed after birth, not fetally. These callosities are often not well-developed even when present.

The infrequency of these callused pads may well be related to the fact that pongids are nest builders; these nests serve as resting and sleeping platforms. Natural selection in this case favored a behavioral rather than a morphological trait.

*Fig. 7–8*   The external sexual dimorphism of the female and male orang-utan (*Pongo pygmaeus*). Note the round curvature to the top of the male's head due to the crown pad of the sagittal crest. (San Diego Zoo Photo by Ron Garrison.)

## THE ORANG-UTAN

Once an inhabitant of mainland China, the orang-utan is now found only in the tropical forests of Borneo and northern Sumatra. The orang-utan is the only Asiatic great ape. This frugivorous pongid is almost completely arboreal yet displays extreme sexual dimorphism (Fig. 7–8). This is an exception to the idea that sexual dimorphism is a terrestrial morphological adaptation complex. Grown orang-utan males are approximately twice the size of the females; their weight is also close to double the female's. In overfed, inactive captivity situations, the male can weigh over 350 pounds and the female about 180 pounds (Fig. 7–9). The male usually has a well-developed sagittal crest, which adds artificially to the height of his head (Fig. 7–10). He also possesses large, flaring cheek pads that contain subcutaneous fat and connective tissue. These pads give the face an exceedingly broad appearance. Hanging like an apron from the male orang-utan's neck is a huge, pendulous laryngeal air sac, which is likely used as a vocalization chamber: the female has a far smaller air sac. The male also has scent glands located in the center of his chest; the female lacks these. The reasons for such gross sexual selection are not known.

*Fig 7–9* The obesity caused by captivity. Note the massive cheek pads and laryngeal sac of the male orang-utan. (Photo by Peter Simon.)

*Fig. 7–10*  The skeleton of the male orang-utan. Note the sagittal crest atop the skull and its meeting with the occipital crest on the back of the skull. (Courtesy of the American Museum of Natural History.)

A brachiator by anatomy, the adult orang-utan is a very cautious slow-mover who will often use quadrupedal movements more than brachiation because of his large size and heavy weight. Schultz's work has shown that as in the case of gibbons, the incidence of healed fractures is great in the orang-utan. His clumsy falls can be of considerable distances in the dense tropical forest; the density of the forest probably helps to keep down fatalities from falls. Anatomically this creature is designed for excellent brachiation, as is evidenced in young individuals.

Due to the lack of the ligament (ligamentum teres) which connects the femur (thigh bone) to the hip socket, and the thigh-muscle arrangement, the orang-utan is able to move his lower limbs in planes similar to his upper ones. This latter trait, coupled with his very prehensile hands and feet, make the orang-utan the true quadrumanous (four-handed) primate. Quite long upper limbs, long, broad hands, and a very small thumb are part of his brachiator's anatomy. The orang-utan's foot is an approximation of his hand; parallel reduction has occurred in the big toe to the extent that very often one bony element of this toe is absent. The bones of the individual digits are very curved, a hook-grip adaptation.

When on the ground, the orang-utan can stand semi-erect but not too easily; some bipedalism also is possible. In quadrupedal position and movement on the ground, the orang-utan is a **fist-walker**. In the hindlimb, the orang-utan walks on the outside edges of the foot rather than the sole (plantigrade). Anatomically and functionally the orang-utan is at his best in the trees. His arboreal adaptations are superb.

The hairy coat of the orang-utan is generally a reddish-brown color, although intergrades of color are found. Infants are often bright brick-red. The upper body hair is rather unusual in that it appears to continually grow, often to extreme lengths. The skin color is usually dark with some "chocolate" blotching. The anatomical differences between the Bornean and Sumatran orang-utans appear to be minor. Notably the Borneo subspecies has a bareface, while the Sumatran one has a beard.

In captivity, the orang-utan is most often described as a lethargic and unresponsive primate. Those who have worked with these fascinating creatures often find them to be inquisitive and rather friendly. Most large adult primates in zoos become apathetic due to the unnaturalness and limited territory of their cages. We have no reason to believe the orang-utan is any less intelligent or evolved than the chimpanzee or gorilla.

Approximately 10 percent of all nonhuman primates are vanishing species. One of the most endangered of all mammals is the orang-utan. There are probably less than several thousand left in the wild (it is possible to go for days in the rain forests of Borneo without seeing one). The orang-utan has been disappearing because of the folly and greed of man. Always a prized zoo animal, the orang-utan has brought sums of well over 5,000 dollars per capture. Collectors and trappers always prefer to capture young orang-utans because of the lack of danger they present, space they take up, and food they consume. The mother is often killed when the young orang-utan is captured, thus eliminating the most important reproductive element. Most zoos will now only accept orang-utans from other zoos. Capture and trade have been outlawed in their native territories, yet poaching still goes on. Natives also kill them for their meat and for trophy skulls. The native government is tearing down jungles in order to enter the logging industry; this is cutting away at the habitats of the orang-utan, causing crowded conditions. Fortunately, there are now approximately 500 orang-utan in zoos and primate centers around the world. In recent years, captivity breeding programs have begun to be successful (Fig. 7–11). We will likely have to rely on these programs to save the species; although conservation measures are being taken, they are on a very small scale. We have no guarantee that a naturally selected type of orang-utan will be secure for the future. Captivity breedings are unnatural and should not be expected to produce biologically

*Fig. 7–11* Twin orang-utan infants at six months of age. (Courtesy of L. Heck and Tierpark Hellabrunn, photo by T. Angermayer.)

natural animals. Crowded conditions in the wild may drastically alter normal breeding patterns.

Although a number of competent investigators have observed the orang-utan in its natural habitats, these investigations have been of only a few weeks' duration. Thanks to the lengthy field studies of John MacKinnon and Dr. David A. Horr, we have a better but still incomplete knowledge of the behavior of the orang-utan.

In zoos orang-utans are grouped into artificial families; in the wild they have a completely different type of social organization. The only permanent unit is the adult female and her offspring. Adult males are solitary animals except when mating. The female-offspring units maintain small home ranges that are fairly permanent. Female-offspring ranges can overlap. Adult males probably have quite large ranges that sometimes overlap with those of the female-offspring. Meetings between female-offspring units and with solitary males are very infrequent. While orang-utans will come down to the ground and spend some time there if the canopy food supply is inadequate, the arboreal environment is the more constant context.

Investigators who have visited the natural habitat of the orang-utan have been impressed with the relative scarcity of these creatures. This has largely been blamed on the predation of man upon them. Horr's study has shown that this is not the complete answer. Females wait until their offspring are about three years old before they return to sexual activity. In some cases this can possibly mean a sexual hiatus of four years—an obvious partial explanation for their low population numbers. In overcrowded conditions, MacKinnon found that females would not mate with males in a natural state; the orang-utan male therefore resorts to unusually aggressive behavior and "rape." Horr's survey of the ecology of the Bornean orang-utan's habitat found their food to be sparse but evenly distributed. To a considerable degree, the orang-utan low-population densities, precise home ranges, and female sexual behavior may all be geared to getting the most out of the available food supply.

The male orang-utan is a difficult creature to assess in the wild. He plays no protective role; his behavior is low keyed, even when reacting to possible threats. Males do vocalize. Their calls are spacing and location devices and also serve for mating. If sexually receptive and in an uncrowded situation, females will respond to these mating calls. Males avoid one another consistently. Horr has suggested the interesting hypothesis that the extreme sexual dimorphism of the orang-utan male is a function of intense male-male competition for females. If time allows, additional studies will hopefully throw some light on this and other facets of the biology of the orang-utan.

### THE CHIMPANZEE

The chimpanzee is the most familar nonhuman primate. Young chimpanzees are known for their attention-seeking antics and imitative behavior. The adult chimpanzee is less extreme in behavior and in captivity tends to be somewhat apathetic. Most people think of chimpanzees as they are when young animals. The adult chimpanzee is a formidable creature. Of orang-utan size, but weighing less, adult chimpanzees, especially males, can be over five feet in height and weigh over 100 pounds. Sexual dimorphism is not extreme in chimpanzees. While males are generally larger and heavier, the differences are not of the magnitude of the orang-utan and gorilla sexual dimorphisms (Fig. 7–12).

*Fig. 7–12* An adult male chimpanzee. (Courtesy of A. H. Schultz and the Carnegie Institution of Washington.)

In habitat, chimpanzees are forest animals but will spend time in savannah grassland and woodland areas. The chimpanzee is arboreal three-quarters of the time. He builds a new nest each evening to sleep in; these nests are high above the ground. In diet, he is mainly a vegetarian, a fruit-eater, but will occasionally consume animal protein in the form of termites, bush pigs, red colobus monkeys, and even an occasional very young or sick baboon.

Anatomically the chimpanzee is a brachiator. In locomotor pattern, he will brachiate over small distances. Typical of great apes, the younger chimpanzees are the most proficient brachiators due to their lighter body weights. Quadrupedalism is common. When on the ground, the quadrupedal posture is most often preferred; but when in high grass, semierect postures and bipedal locomotion are sometimes employed. When walking quadrupedally, the chimpanzee's hand rests on the middle segments of each finger, as does the gorilla's; thus we call these primates **knuckle-walkers** (Fig. 7–13).

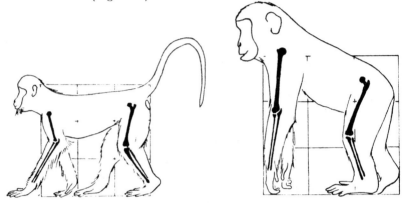

*Fig. 7–13*  Comparison of the extremity proportions of a digitigrade terrestrial quadruped (Old World monkey) and a pongid, the chimpanzee (right). Note the knuckle-walking position of the chimpanzee and the level and inclination of the upper torso. (After Ankel, 1970.)

Typical of brachiators, the chimpanzee's hindlimbs are relatively short compared to his trunk and forelimbs. His short thumb makes a true precision grip impossible. The chimpanzee and gorilla hand anatomies are designed for hook grips to the extent that the tendons of the flexor muscles of the forearm are so short that when the wrist is bent, a hook grip is formed. This is responsible for the knuckle-walking posture and also the fist-walking of the orang-utan. Thus a digitigrade (on palms and finger tips) posture is impossible in these great apes. The big toe is extremely divergent (abducted) from the other digits, thus making the foot an excellent grasping organ.

Ischial callosities occur in approximately 35 percent of chimpanzees but are often not well developed. Females show an unusually large periodic sexual swelling, while orang-utan females have a modest swelling and only during pregnancy.

Relatively, the orang-utan has the smallest and flattest ears of all the primates. The chimpanzee, however, is at the other extreme; his ears are frequently described as "jug handles." The body hair is usually black, but some light-haired animals have been reported. There is a tendency for premature baldness in chimpanzees, especially the females.

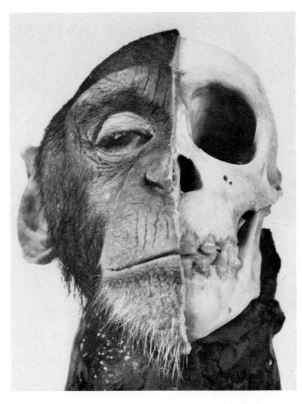

*Fig. 7–14* A cutaway dissection of a young male chimpanzee (*Pan troglodytes*). (S. I. Rosen.)

The supraorbital regions of the chimpanzee are usually in the form of well-developed ridges (rather than tori), although there is variability here. Sagittal crests are not usually found in the male or female. Pigment of the face is variable: some are dark, some pale, and others have a mottled appearance.

In an area south of the Congo River lives a variant of the typical chimpanzee known as the pygmy chimpanzee, or bonobo (Fig. 7–15). Though small in size, the pygmy chimp is not different anatomically from other chimps except for some webbing between the second and third toes and less body hair. This animal has been placed in a separate species due to its assumed reproductive isolation. Anatomically, it can be considered only a subspecies of chimpanzee.

Many claims have been made for the chimpanzee's unusual intelligence, especially in the recent work on teaching this animal symbolic languages. The other great apes are probably as intelligent. The chimpanzee is easier to test than the orang-utan and gorilla because of his personality, imitative behavior, and attention seeking. Some monkeys may in fact be as intelligent as chimpanzees.

*Fig. 7–15* The pygmy chimpanzee, or bonobo (*Pan paniscus*). (Tierbilder Okapia, Frankfurt-on-Main. Photo by B. Grzimek.)

The behavior of the chimpanzee is a topic of interest to all because these nonhuman primates appear so often to mirror human behavior. Although a number of students of animal behavior have studied chimpanzees in the wild, the fieldwork of Vernon Reynolds and of Jane van Lawick-Goodall form the major contributions to our knowledge.

The wild chimpanzees are "organized" into extreme fusion-fission societies. They continually group, disband, and form into new groups with different compositions. Theirs is a completely open society, a rarity among nonhuman primates. Social units range from solitary chimpanzees to groups of eighty individuals. The only unit with any stability is the adult female-offspring one. Evidence has been gathered which indicates that this maternal unit has some permanence in a few cases. Mature offspring may prefer to be around their mother and sometimes return to be near her when both are well into maturity. Living in forest and savannah, a chimpanzee may forage up to eight miles a day, but individuals have no real home ranges. The population as a whole has such a range.

Chimpanzee society is quite friendly. At locales where food is abundant, chimps will gather together and have "food carnivals," drumming on tree trunks to signal one another of food location. Greetings are almost humanlike, especially among chimpanzees who know one another and who have a mutual attraction. Hugs, kisses, and pats are common in such settings. Chimpanzees also exhibit the unusual nonhuman primate quality of cooperation. They will engage in cooperative hunting behavior to obtain fresh meat. In response to food begging, chimps will share a portion of their food with one another in some cases. Although dominance hierarchies do exist, they are hardly evident. Submissive behavior can consist of "presenting" the rump to the more dominant chimp; this is found in semiterrestrial monkeys also. Touching, especially the genitals, bowing, and crouching (extreme presenting) are also part of the submissive repertoire. On a general level, chimpanzee societies can be described as open, tolerant, and nonaggressive.

Van Lawick-Goodall's studies changed to some extent our concept of man as "the tool-maker." She discovered that chimpanzees will strip grass stalks and use them and sticks as effective probes for fishing termites out of the ground. This is tool-modification for an immediate purpose. These "tools" are not used for future fishing tasks. Interestingly, one European zoo keeper claimed he saw this behavior in zoo chimps decades ago; unfortunately he reported it too late. Van Lawick-Goodall also reported chimpanzees mashing leaves in their mouths and later using these leaves as sponges to obtain drinking water. In experiments with dummy predators, Kortlandt found that savannah chimps will resort to using sticks as clubs and will attack a potential predator. These reports of "toolmaking," and of tool and weapon use indicate that the chimpanzee may be an excellent theoretical model for determining the fossil behavior of our earliest ancestors. The study of the pygmy chimpanzee may be especially fruitful because of its close proximity to the size of the smaller type of australopithecine and dryopithecine.

## THE GORILLA

Perhaps the most maligned of all animals has been the gorilla. Regarded in blind ignorance as a fiercely aggressive hostile beast having an unusually ravenous sexual appetite, the gorilla in the wild and in captivity is a quite different animal from its myths. Primatologists have often described this largest of all living primates in more human terms than any other primate—aloof, shy, gentle, stoic, and introverted. Through field study, notably that of Dr. George Schaller, we find the gorilla in the wild to be a gentle creature, and many have found this true in captivity settings. The sex life of the gorilla is probably the least interesting and certainly the least frequent of all primates. Those who have had personal experience with gorillas, especially young ones, acquire a fondness for them that exceeds the usual human relationship with pets.

The gorilla is more confined in habitat than the chimpanzee. The gorilla populations are limited entirely to tropical rain forests. A forest floor dweller, the gorilla has a diet of bulky, tough vegetation such as vines, foilage, bamboo, and other high-cellulose food. The large, conical, sexually dimorphic, projecting canines of this creature are used primarily for stripping tough vegetation rather than defense. Even more than the chimpanzee, the greater part of its waking day is spent in eating—up to eight hours a day. Rest and sleep are taken usually in ground nests or simple tree nests close to the ground. Sunbathing is a frequent pastime.

Sexual dimorphism is very evident in the gorilla. Males can occasionally weigh over 400 pounds in the wild and even up to 600 pounds in inactive captivity; females can weigh up to 250 pounds but are usually much lighter. An adult male can exceed six feet in height and have an arm span of up to ten feet. While the strength of large adult male chimpanzees has been compared to that of five grown men, that of the male gorilla is unfathomable. The male quite often has a massive sagittal crest which appears to grow through adult life, as long as the dentition is functional. The sagittal crest is well padded by subcutaneous connective tissue on each side. This crown pad gives the living male cranial vault a misleadingly filled-out appearance. The mature male gorilla is easily identifiable by the silver-grey hairs on his lower back. Because of a somewhat saddle-shaped appearance to this silver hair patch, males are sometimes referred to as silver saddle-backed males. Greying will also occur over other body areas in both males and females. The adult male also tends to lose chest hair.

*Fig. 7–16* The western lowland gorilla (*Gorilla gorilla gorilla*). Note the silver-gray of this adult male. (San Diego Zoo Photo by Ron Garrison.)

While only young gorillas are functionally capable of brachiation, and do so, the gorilla has the anatomy of a structural brachiator. While they are also capable of erect posture and bipedal locomotion, these are rather infrequent events. The gorilla, like the chimpanzee, is best described as a quadrupedal knuckle-walker.

The pongids are notable in having extensively well-developed laryngeal air-sac systems. While the orang-utan's is mostly external, the air-sac systems of the chimpanzee and gorilla are internal. It is likely that when the gorilla cups its hands and "beats" its chest, the air-sac system is inflated, creating a resonating chamber and a pillowlike protection for the rib cage. It should be noted that other nonhuman primates such as baboons and macaques have laryngeal air sacs; even vestigial remnants are seen in some humans. There may be a correlation between diet and air-sac size; it seems the more herbivorous the species, the larger the air sac. Vocalization habits are involved also.

Three geographic variants (races) of gorilla seem to exist; a western lowland gorilla, an eastern lowland, and a mountain (eastern lowland) gorilla. The eastern lowland gorilla seems to be morphologically intermediate between the western and mountain gorillas. The mountain gorilla has a notably blacker coat and longer fur (Fig. 7–17). The western lowland gorilla sometimes has rust-colored hair atop its head. The mountain gorilla has larger teeth and jaws and a foot that more closely approaches man's than any other pongid; its big toe has less opposability than that of any other ape. The lowland forms appear to have proportionately longer limbs and narrower heads than the mountain variant. Eastern and western gorilla populations are geographically separated from one another so that they are probably isolated reproductively. There have been reports of pygmy gorillas and hybrid chimpanzee-gorilla crosses, but none have so far been documented.

Unfortunately the gorilla is becoming a vanishing primate. Although the gorilla is possibly the real "king of the jungle," and for all purposes has no natural enemies (although there have been a few reports of leopards killing gorillas), he does have one "unnatural" enemy—man. Mercenaries in the past have killed gorillas for sport. Agriculturalists kill them for their crop-raiding activities and take their feeding areas for agricultural expansion. Natives kill them out of fear, and animal collectors occasionally kill adults to obtain the smaller, younger gorillas. Fortunately, the gorilla is not yet extinct. With proper conservation procedures, the gorilla may be saved. Even with the recent successes of gorilla-breeding in captivity, there is some question as to whether we get a biologically normal gorilla from this; likely we do not.

In recent years, several primate biologists have taken the position that the gorilla and chimpanzee are not very distinct from each other

*Fig. 7–17* An adult male mountain gorilla (*Gorilla gorilla beringei*). Note the effect of the sagittal crest and its padding. (San Diego Zoo Photo.)

and should be grouped together in the same genus, *Pan,* which is the current taxon of the chimpanzee. While many similarities of anatomy and biochemistry are cited as reasons for this "lumping," significant differences between these giant primates are ignored. To place both in the same genus would indicate that the chimpanzee and gorilla are more similar than they are. The creation of a "supergenus" would be meaningless. If this single-genus classification gains acceptance, man must be placed in the genus *Pan* also because of the nature of the evidence used for support.

The behavior and social organization of the gorilla is in sharp contrast to that of the chimpanzee and orang-utan. Our knowledge of gorilla behavior is owed largely to Dr. Schaller's study of the mountain gorilla. Gorillas live in groups varying in size from three to thirty animals. They maintain low-population densities and relatively large home ranges compared to monkeys. As in the case of the chimpanzee, the home range concept is applicable only at the population level. Every gorilla group has at least one silver-backed male and several females. The silver-backed male is dominant over the black-backed males and females. The presence of other silver-backed males in the group sets up a linear hierarchy based on body size and likely age. Some males are solitary at times and at other times join groups. Basically gorillas do not form fusion-fission societies like the chimpanzees, but rather stable groups. Within the group there is not much social interaction. There is little mutual grooming or body contact. Individual gorilla groups are not generally hostile to one another. There is no competition for food or sexual access to females. The nucleus of each group is the dominant silver-backed male and the females.

Dominance is by gesture not aggression. Most dominance inter-actions take place on trails. Submissive behavior usually takes the form of giving the right of way to the more dominant gorilla. Chest beating can be used as an aggressive gesture to irritate other gorillas, or as

*Fig. 7–18*   The world's only known white gorilla "Snowflake." Technically this western lowland gorilla is not a true albino. (Courtesy of J. Sabater Pi.)

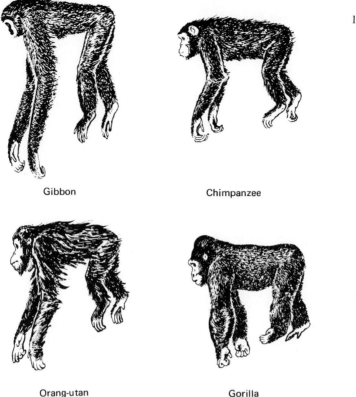

Gibbon          Chimpanzee

Orang-utan          Gorilla

*Fig. 7–19*   Relative body proportions of the four anthropoid apes. (Courtesy of
G. E. Erickson and the Zoological Society of London.)

emotional release. Vocalizations are low keyed and are probably not used
too much for communication. Bluff charges are employed in some dominance behavior. In many ways the behavior of the gorilla is more like
that of an arboreal monkey species than a semiterrestrial primate. Such
behavior may indicate that gorillas were primarily arboreal creatures in
the evolutionary past.

The pongids present us with unique opportunities of studying
species whose anatomies and other biology are remarkably similar to
our own. Their use as animal models for human disease has barely begun, yet this holds significant promise for biomedical research.

As a grade of primate organization, they are a fairly logical expression of one end of the range of nonhuman primate variation. We are
attracted to studying them because of an emotional factor: they are
anthropomorphic (manlike). Men have even been known to wear "gorilla
suits" on special occasions. In the next chapter we will see how closely
they do in fact resemble man.

# SUMMARY

*Hominoidea*
*Hylobatidae*
*Pongidae*
*Hominidae*
COMMON GIBBON
SIAMANG GIBBON
THE HYLOBATID PATTERN
    Full-Time Brachiator
    Tailless
    Forelimbs Extremely Long
    Frugivorous
    Wholly Arboreal
    Monkey-Sized
    Ischial Callosities Always Present
    Little Sexual Dimorphism
ORANG-UTAN
CHIMPANZEE
GORILLA
THE PONGID PATTERN
    Largest and Heaviest Nonhuman Primates
    Vegetarians
    Largely Frugivorous
    Large, Projecting Canines
    Diastema
    Separate Molar Cusps
    Massive and Prognathous Jaws
    U-shaped Dental Arcade
    No Chin
    Simian Shelf
    Extensive Cheek Bones
    Sagittal Crest
    Massively Developed Supraorbital Regions
    Lack a Forehead
    Prolonged Semierect Postures and Bipedal Locomotion
    Large Occipital Crests
    Quadrupedal Pelvis
    Tailless
    Very Large and Complicated Brains
    Structural Brachiators
    Relatively Long Forelimbs
    Reduced Thumbs
    Poor Precision Grips
    Prolonged Life Spans
    Delayed Maturity
    Considerable Sexual Dimorphism
    Ischial Callosities Infrequent
FIST-WALKERS
KNUCKLE-WALKERS

# Chapter 8

# Man

Several years ago, W. E. LeGros Clark correctly pointed out that the terms "man" and "human" have become biologically meaningless words, filled with emotional rather than scientific content. Yet it is difficult not to use these words when speaking of the biology and evolution of man. These terms are used in this text, but we are not speaking of man the philosopher, or man the dreamer. The subject is man the animal—a primate.

### The Hominid Pattern

Man, more properly *Homo sapiens sapiens*, is the only living member of the primate family Hominidae, the hominids. This family does include fossil hominids, which we will consider in the second part of this book. Naturally, it is difficult to look at oneself as one looks at other animals. What separates man from the other living primates is his culture, a form of **nonbiological adaptation**. Thus, man the philosopher is the unique primate. There are many good anthropology texts that deal with the cultures of man; the student is encouraged to investigate this meaningful facet of man the primate.

Man is a **giant primate** who has the widest distribution of all living members of the order *Primates*. Many aspects of this primate's culture such as clothing, use of fire, and transportation have allowed him to live in, and at least visit, almost every locality on this planet. This wide geographic distribution does not rest only on nonbiological adaptation; hominids are **extremely generalized** primates in anatomy, except for the central nervous system. The hominid primate deviates less from the primate pattern than any other living primate. His anatomy is so generalized that he can fit into a great variety of econiches. This biological plasticity has insured the survival of the one living species of hominid.

Morphologically, the hominid occupies a logical point at one end of the primate spectrum. The anatomical traits that many authors cite as being unique to man are, in fact, trends common to a large number of the primates, especially the higher ones. The hominid is to a certain extent an exaggeration of the primate pattern. Today there is only one species of hominid, an eloquent testimony to precise, selective, evolutionary processes working in the past for a certain type of anthropoid, man. Compared to the other living primates, man exhibits only minor and insignificant variations.

The **brain** of man is **very large and complex**. It is an exaggerated hominoid brain (Fig. 8–1). The outstanding neurological hallmark of man is the extremely convoluted cerebral cortex, where nerve fibers enter and leave the brain, and important biochemical-association functions take place. Both absolute and relatively, the hominid brain is almost too large. It ranges from 1,000 to some 2,000 cc. in size, with the average European male brain at 1,450 cc. (Table 8–1). The British anatomist and physical anthropologist Sir Arthur Keith once suggested that there was a "cerebral rubicon" for man—that is a brain size where man begins.

**Table 8–1.**    Average Cranial Capacities (in cm³) of Adult Male Hominoids

|  |  |
|---|---|
| Gibbon | 104.0 |
| Chimpanzee | 390.0 |
| Orang-utan | 425.0 |
| Gorilla | 525.0 |
| Man | 1,450.0 |

He placed this point at 750 cc. However, Schultz calculated the cranial capacity of a large male gorilla to be 752 cc. and showed the fallacy of such anatomical demarcation points.

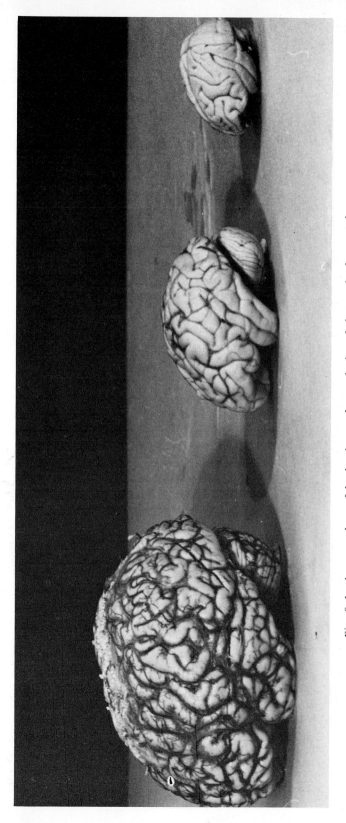

*Fig. 8–1* A comparison of brain sizes and complexity of the cerebral cortex in (left to right) man, the chimpanzee, and an Old World monkey (rhesus.) (S. I. Rosen.)

The expansion of the hominid brain, especially in the frontal region, has brought about significant changes in the hominid skull. Only man has a **true forehead**. The supraorbital region is relatively undeveloped; there are no massive ridges or tori. The face is under the brain, not just in front of it. The widest part of the skull is up high, toward the back (parietal bone region), and not low near the skull base as in the pongids.

Some of these cranial features are also related to the hominid dentition. Man is probably best described as a **complete omnivore**. He is so omnivorous that he has even been known to eat members of his own species. Although he has the catarrhine dental formula, his **dentition is reduced** in size. **No large projecting canines** are found, nor is a diastema commonly present. Although the canine teeth have long roots, they are small and shaped like the incisors, which are also small. Man has a **balanced dentition**—no teeth are exaggerated in size. The teeth are proportionate to one another. Tooth rows are arranged into a **parabolic shaped dental arcade** (semicircle), unlike the U-shaped arcades of pongids. The hominid molar tooth has low, blunt cusps in contrast to the high, sharp cusps of the pongid molar. The dental apparatus is supported by **relatively reduced jaws** compared to the pongids. Because the temporal muscles are not large and the braincase is greatly expanded, sagittal crests are not found. The lower jaw, or mandible, is buttressed in front by a **true chin** (mental eminence) rather than the simian shelf of the pongids and some monkeys (Fig. 8–2).

In profile, the hominid face bears a distinct nasal bridge instead of the depressed one on the "dished-out" pongid face. The remainder of the skull **lacks heavy cranial muscle markings** except for an ex-

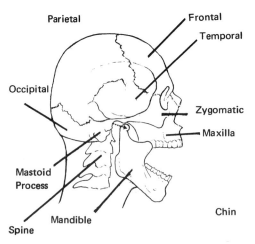

*Fig. 8–2* The human skull. Note the major cranial bones and how they relate to one another. (After Du Brul, 1963, by permission of the Biological Sciences Curriculum Study.)

tremely **well-developed mastoid process** which serves as a point of origin for important musculature to the front of the neck; this process is generally undeveloped in most nonhuman primates except for the gorilla, where it is a variable trait. Usually **heavy neck musculature is lacking** because the hominid skull is balanced on the spine for erect posture and requires no overdeveloped musculature to hold the head up. This

**Table 8–2.** Basic Primate Locomotor Groups

| Locomotor Group | Body size | Body features | Primate |
|---|---|---|---|
| *Vertical clinger and leaper* | small, medium, and large | long hindlimbs and long trunks | indris galago tarsier |
| *Springer* | small | rather generalized, slender trunks, short limbs but hindlimbs longer | marmoset owl monkey titi |
| *Climber* | medium | moderately long limbs, hindlimbs longer, trunks relatively shorter than springer's | cebid monkey many macaques guenon |
| *Terrestrial quadruped* | large | long limbs of approx. equal length, trunk proportionate to limb length | baboon some macaques patas monkey |
| *Brachiator* | large | forelimbs longer than hindlimbs (often exaggeratedly so), trunks relatively short compared to limbs, special modifications in hands, prehensile tails in some | Grade 1—colobine<br><br>Grade 2—howler and woolly monkey<br><br>Grade 3—chimpanzee gorilla<br>Grade 4—spider and woolly spider monkey<br>Grade 5—gibbon orang-utan |
| *Habitual biped striding bipedalism* | large | hindlimbs longer than forelimbs, long trunk | man |

is further reflected in the approximately **centrally located foramen magnum** and occipital condyles.

The postcranial ("behind the head") anatomy of the modern hominid is distinct from other primates in features that reflect his **fully erect posture** and **habitual bipedal locomotion**. Dr. John Napier has properly described this locomotor pattern as **striding bipedalism** (Table 8–2). This pattern begins with a kickoff on the toes, especially the big toe, of one foot aided by the large extensor muscles of the hominid hip and thigh. There is then a transfer of weight to the unique hominid heel; then to the ball of the foot; and finally to the toes again. To accommodate the body's changing center of gravity and to allow more area for the extensor muscles to originate from, the upper portion, or **iliac portion** (blade), of the **pelvis** has been **greatly broadened and shortened** in man (Fig. 8–3). The spine has acquired secondary curvatures that give it an S-shape in side view. The lower limbs are relatively and absolutely long compared to the upper ones, and the knee and thigh joints are capable of full extension. While man lacks ischial callosities, it is interesting that when man is in a sitting position, the majority of his body weight does fall on the ischial portion of the pelvis. Unlike other primates, man is rather overpadded in this area by a large gluteus maximus muscle and abundant fat and connective tissue.

The hominid foot has changed radically from the primate pattern (Fig. 8–4). While all the muscles for prehensibility are still present, it is a relatively **nonprehensile foot**. Some authorities believe that next to

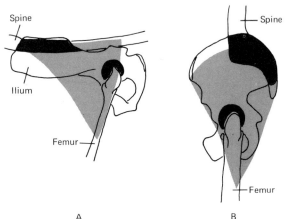

A                                    B

*Fig. 8–3*  A comparison of a quadrupedal monkey (*A*) and the bipedal human. The upper darkened area represents the articulation area between the spine and pelvis (sacroiliac joint). The lower darkened area is the hip socket. The shadowed triangular area represents the extent of the gluteus maximus muscle. (After Ankel, 1970.)

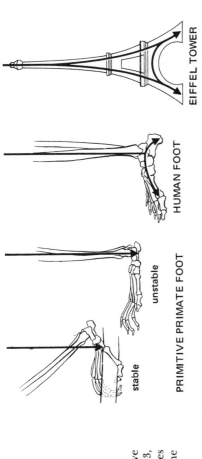

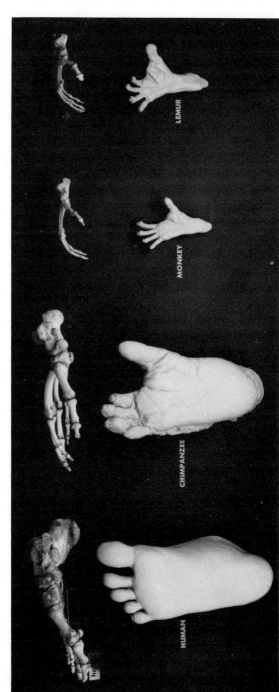

*Fig. 8-4* A comparison of representative primate feet. (Above—after Du Brul, 1963, by permission of the Biological Sciences Curriculum Study; below—courtesy of the American Museum of Natural History.)

EIFFEL TOWER

HUMAN FOOT

PRIMITIVE PRIMATE FOOT

unstable

stable

LEMUR

MONKEY

CHIMPANZEE

HUMAN

the central nervous system, the foot of man is his most specialized and highly evolved anatomical area. The pelvis, in my opinion, is even more specialized than the foot.

Needless to say, this erect posture has freed the human forelimbs. As a secondary result of this, man has a long and widely divergent thumb (Fig. 8–5). This anatomy enables him to have a **true precision grip** (Table 8–3). When quadrupedal as an infant, the hominid hand functions in a digitigrade posture, not in the knuckle-and fist-walking postures of the pongids.

While the living hominid is a **terrestrial** primate, he exhibits only **moderate sexual dimorphism**. Differences in facial and body hair are exhibited between male and female hominids, but to a lesser degree than in the other primates. There are **no sexual swellings** in either males or females. Not fully explainable are the large, pendulous mammary glands of the female hominid. Man also has a scant amount of body hair, reflecting a general trend in the higher primates.

In most features of the primate pattern, man represents a logical end point. For example, he has an **extremely prolonged dependence period** and **extremely delayed maturity**. These traits are likely the re-

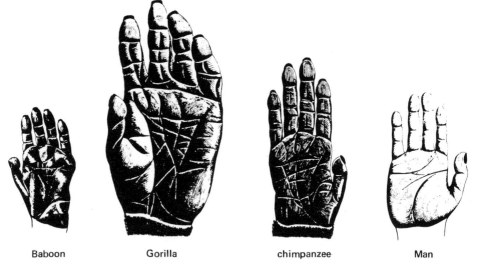

| Baboon | Gorilla | chimpanzee | Man |

*Fig. 8–5* Anthropoid hands. Compare the relative lengths of the thumbs and their relationships to the total hand lengths. (After Biegert, reprinted from Sherwood L. Washburn, ed. *Classification* and *Human Evolution* [Chicago: Aldine Publishing Company, 1963]. Copyright © 1963 by the Wenner-Gren Foundation for Anthropological Research, Inc. Reprinted by permission of the author and Aldine-Atherton, Inc.)

**Table 8–3.**     Important Primate Trends

| | Opposability of digits | Precision grip* | Bipedalism |
|---|---|---|---|
| Arboreal quadruped | poor | none | none to poor |
| Terrestrial quadruped | excellent | excellent | good |
| New World brachiator | poor | none | good |
| Old World brachiator | good | poor (except gibbons) | good |
| Man | excellent | excellent | excellent |

*Precision grip here is measured by the degree to which the soft digital pad of the thumb can be brought in opposition to the pads of the other digits. This is the so-called "pulp-on-pulp" opposition. Precision grip here is not a measurement of manual (especially digit) dexterity.

**Table 8–4.**     Comparative Cranial Total Morphological Patterns of Pongids and Hominids

| Trait | Pongids | Hominids |
|---|---|---|
| Forehead | absent | well-developed |
| Supraorbital region | heavy browridges, or tori | poor development |
| Cranial vault | playtycephalic (flat) | high and rounded contour |
| Occipital region | high and extensive nuchal cresting | low and poorly marked |
| Foramen magnum and Occipital condyles | posterior position | almost in center of skull base |
| Nasal bridge | depressed | high |
| Nasal aperture | large | moderate to small |
| Incisor teeth | large and slightly procumbent | small and vertical |
| Canine teeth | large, conical, and projecting | small, incisoriform, and nonprojecting |
| First premolar teeth | sectorial | nonsectorial, molarized |
| Molar teeth | large with high and sharp cusps | moderate size with low, blunt cusps |
| Diastemata | present | normally not present in adults |
| Dental arcade | U-shaped | parabolic |
| Jaws | massive | moderately small |
| Simian shelf | often present | always absent |
| Chin | absent | present |
| Facial profile | dished-out | relatively even |

sult of selective evolutionary pressures for a biology that would insure the proper socialization and enculturation of the young. In recent history, man has artificially prolonged his own life span. Prehistoric populations had life spans roughly equivalent to the living pongids.

The subspecific variations found in living hominids are not particularly impressive when compared to many of the nonhuman primates. All of these varieties, or races, of man can interbreed with one another and produce fertile living offspring; they are thus not separate species but, at the most, subspecies. Hominid variations have probably come about as rapid and minor physical adaptations to particular climates and geographical settings in the last twenty to thirty thousand years. So much interbreeding has occurred that racial classifications such as Negro, Caucasian, and Mongoloid have become biologically meaningless.

## Brachiator versus Nonbrachiator Origins for Man

Thomas H. Huxley, the overzealous defender of Charles Darwin, is largely responsible for the still popular concept of man being descended directly from a brachiating great ape. An often quoted statement of Huxley's describes his position: " . . . the structural differences which separate Man from the Gorilla and the Chimpanzee are not so great as those which separate the Gorilla from the lower apes (monkeys)." This brachiator school has had such eminent followers as the English anatomist G. Elliot-Smith and Sir Arthur Keith, the German naturalist Ernest Haeckel, and the American vertebrate paleontolgist, William King Gregory.

There have been equally illustrious antibrachiationists. We have already mentioned the English anatomist F. Wood Jones and his tarsier ancestry theory. The French zoologist St. George Mivart, definer of the order *Primates*, believed the hominid ancestor to be a very generalized primate; he also held that numerous cases of parallel evolution at all primate levels largely obscure the precise ancestry of man. The Scottish paleontologist Robert Broom was also an adherent of this school of thought.

The current chief protagonist of the brachiator school is Dr. S. L. Washburn. The modern concept might better be termed the **brachiator-knuckle-walker school**. Until recent years, Washburn and his followers based their conclusions on anatomical traits man shares with the pongids, especially the chimpanzee, while ignoring equally important traits man shares with other primates. Recently, members of this school have claimed that man is a direct descendant of chimpanzees in the Pliocene, five to seven million years ago. This newer brachiator-knuckle-walker position is based on similarities in albumen protein serum chemistry between the modern chimpanzee and man. Constant rates of biochemical evolutionary change are claimed. It is unlikely that such rates can in fact be discovered in terms of evolutionary speed. We know that in terms of gross morphologies, the vertebrates have not evolved at constant rates. Albumen protein is but one facet of primate biochemistry; there are other biochemical factors in which man equally differs from the chimpanzee. Since the first recognizable fossil candidates for human evolution were savannah grassland dwellers like some of the chimpanzees are today, it is only to be expected that similar biochemical and morphological adaptations would take place. The important question is why we differ at all biochemically from the chimpanzee or any of the other apes. The evidence would appear to favor hominids and pongids developing in parallel from a common pre-Pliocene ancestor of some sort.

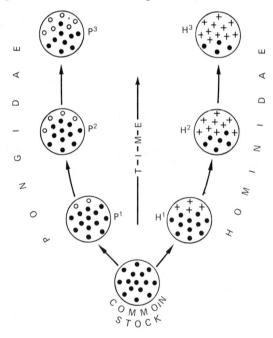

*Fig. 8–6* Diagram representing the divergence of two evolutionary sequences, the Pongidae (great apes) and the Hominidae (modern and extinct man.) The two sequences inherit a common ancestry—characters of common inheritance (black circles). As the lines diverge, each one acquires its own distinctive features or characters of independent acquisition, those distinctive of the hominid sequence of evolution are represented by crosses and those of the pongid sequence by white circles. (After LeGros Clark, 1959.)

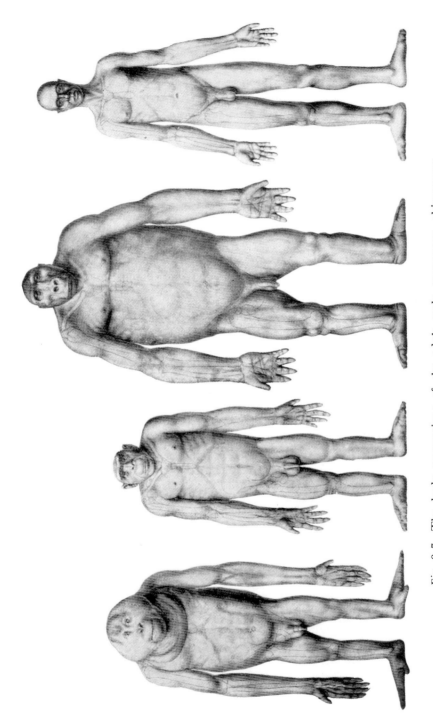

*Fig. 8-7* The body proportions of the adult male orang-utan, chimpanzee, gorilla, and man. The lower limbs are unnaturally straightened and the bodies devoid of hair. (Courtesy of A. H. Schultz.)

The greatest weakness of the brachiator-knuckle-walker school is that the fossil evidence just does not support this thesis.

The leading modern antibrachiationist has been Dr. William L. Straus, Jr. Through comparative anatomical studies, Straus has outlined many similarities between catarrhine monkeys and hominids. This school believes it is more likely that man was derived from some kind of generalized quadrupedal monkey. This type of monkey could have been ancestral to both the pongids and man. Even with Straus's impressive work, the **antibrachiation school** has had almost no following. However, I tend generally to adhere to this line of thought.

The most popular trend is one which could be termed the **pre-brachiation school**. Chief member of this school of thought was **W. E. LeGros Clark**. This school believes that it is more probable that man came from a group of primates that had not become fully adapted for brachiation, yet probably did brachiate at times. This type of primate would not likely be describable as entirely ape or monkey. The dryopithecines largely meet the description of this ancestral group (see Chapter Ten).

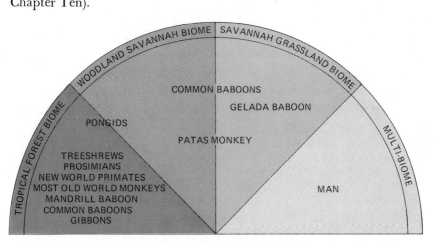

*Fig. 8–8*    The ecological pathways of the living primates.

In the second part of this book, we will look at the fossil primates, especially those who may have been ancestors, at least structural ones, of modern man. The student will be presented with a confusing mass of names and ideas. Like parents who give birth to a child, the finder of a hominid fossil very often adopts this bit of rock as his own child, which he believes to be unique and deserving of a new name. Too many such fossil hunters, skilled and unskilled, have avoided taxonomic convention and have immediately christened their new find with a new and invalid ·

taxon lacking adequate documentation. We find too many synonyms for one type of fossil primate. The introductory student and sometimes the professional also is left in a quagmire of inappropriate taxons and colloquial names. Again, the student should think in terms of total morphological *patterns* rather than a specific fossil primate trait.

## SUMMARY

THE HOMINID PATTERN
 Nonbiological Adaptation
 Giant Primate
 Extremely Generalized
 Very Large and Complex Brain
 True Forehead.
 Complete Omnivore
 Reduced Dentition
 No Large Projecting Canines
 Balanced Dentition
 Parabolic-Shaped Dental Arcade
 Relatively Reduced Jaws
 True Chin
 Lack Heavy Cranial Muscle Markings
 Extremely Well-Developed Mastoid Process
 Heavy Neck Musculature Lacking
 Centrally Located Foramen Magnum
 Fully Erect Posture
 Habitual Bipedal Locomotion
 Striding Bipedalism
 Iliac Portion of Pelvis Greatly Broadened and Shortened
 Nonprehensile Foot
 True Precision Grip
 Terrestrial
 Moderate Sexual Dimorphism
 No Sexual Swellings
 Extremely Prolonged Dependence Period
 Extremely Delayed Maturity
BRACHIATOR-KNUCKLE-WALKER SCHOOL

ANTIBRACHIATION SCHOOL

PREBRACHIATION SCHOOL

# PART II

# THE FOSSIL PRIMATES

# Chapter 9

# From Bone to Stone

*the earliest fossil primates*

While fossilization can be only a simple imprint of the external morphology of a plant or animal in what is now stone, the type of fossilization we are concerned with is the transformation of bone into stone. Bone is approximately 65 percent inorganic matter (mineral salts) and 35 percent organic (perishable living matter). When an animal dies, all organic matter begins a process of self-destruction (autolysis). The inorganic portion remains as a skeleton. Even this relatively hard skeleton is vulnerable to elements of nature such as ground water, humidity, soil acids, microorganisms, and the various animals that chew and eat bone. Under a few favorable situations, this bone can be preserved by further mineralization termed fossilization. Such conditions as percolating limestone, volcanic ash, and highly mineralized ground water can bring new minerals into contact with the bone. The bone can chemically bind these external minerals into its microstructure. The entire process is quite complex and not fully understood. The results are fossils, which are meager but important keys to the animal life of the past.

Since the majority of primates of the past and present have been tropical and semitropical forest dwellers, we should not have many primate fossils because of the damp, acid floors of these environments; yet we do have a surprisingly good number of primate fossils. We do not have by any means an oversupply, nor do we have compartively few as in the case of fossil sharks. The skeletal parts that are most likely to remain long enough to be fossilized are the teeth and the thick, bony jaws. It will become apparent in the coming chapters that there are more fossil primate teeth and jaws than any other remains, and even these are often fragmentary. The other cranial bones have a fair record of preservation, but too often they are too badly crushed to make reasonable reconstructions of the skulls. Very often authorities have major disagreements on whether skulls have been put back together correctly. The braincase (neurocranium) is the major area of disagreement, since differences in reconstruction can significantly alter estimates of cranial capacity.

Probably the most confusing area in primate paleontology is the earliest fossil primates. Many of these specimens once thought to be insectivores have been promoted to primate status; and a few have been demoted back. It is likely that many of these early primates are today in museum drawers going unrecognized for what they are—the first primates.

### The Early Primate Pattern

The pattern one would logically expect to find in the earliest fossil primates is of a very anatomically **generalized and small** mammal, **insectivorelike**. The skull would have had a **small braincase**, probably with a **large olfactory development**. The visual area of the brain probably exhibited some expansion, although the **eyes** would have been in a **lateral** position. The **muzzle** would have been **moderately long** while the **dentition** showed some **rodent features. Slender trunks** with **short limbs** and **long tails** would have made up the remaining anatomy. The barely prehensile hands would have ended in claws rather than flattened nails.

This pattern obviously fits the living treeshrew, and thus its status as a living fossil. The emerging primate pattern is found in varying degrees in the earliest recognizable primates. Many of these creatures were probably evolutionary dead ends, and were not directly ancestral to any of the living primates of today. When we deal with living primates, the level of definition and meaningful analysis is the species. In the fossil record only the genus level is meaningful and accurately observable.

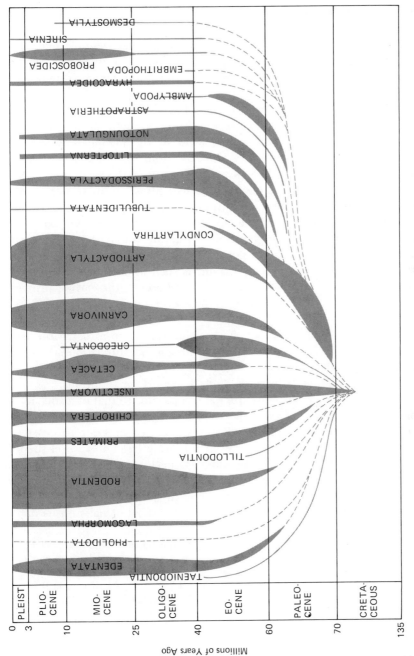

*Fig. 9–1* The time epochs of the last 135 million years and the chronologic distribution of the placental mammals. (After Romer, *Vertebrate Paleontology*, 1966, by permission of the University of Chicago Press.)

157

## The Earliest Fossil Primates

The first primitive mammals appeared during the Mesozoic era, some 180 million years ago. Very late in the Cretaceous period of the Mesozoic, we get the first placental mammals, and among these, the first primates. The best documented early primates are found in the Paleocene (Fig. 9–1). This was a time of widespread tropical and semi-tropical forests, which extended further in longitude than they do today. Obviously the climates were warm. Most of the mammals were small at this time; the primates were no exception. The majority of these mammals were probably like insectivores. Before the beginning of the Paleocene, the insectivores had already had fifty million years of evolution. As is the case today, no naturally occurring placental mammals existed south of Wallace's Line, an arbitrary marker between Southeast Asia and Australia. On the Australian side are found the nonplacental mammals—the marsupials, or pouched mammals such as the kangaroo, and monotremes, or egg-laying mammals such as the duckbill platypus and the spiney anteater. At their peak (the Eocene), the earliest primates were distributed more widely over the world than the nonhuman primates are today.

The fossil mammals that most authorities agree as being the most likely ancestors of the prosimians (although not direct ones) are the **plesiadapines** (*Pleaiadapidae*). These "protoprimates" date from the late and possibly middle Paleocene. They had very small brains and long muzzles (Fig. 9–2). Their body size was comparable to that of a squirrel or small rat. Short-tailed and long-bodied, these creatures were probably terrestrial quadrupeds. They had very specialized anterior (front) teeth which were rodentlike in that the incisors were large and chisel shaped, while the molar teeth were like those of the living lemurs. It has been suggested that these teeth were adapted to a frugivorous diet, but they also would have been adaptable to an insectivore's regime. Evidence further indicates that ***Plesiadapis*** had claws instead of nails on its digits.

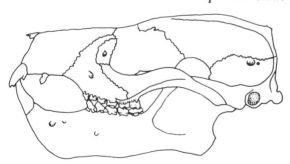

Fig. 9–2  The skull of *Plesiadapis*. Note the lack of a post-orbital bar and plate. The front part of the skull is very rodentlike.

It is doubtful that *Plesiadapis* led to the prosimians, but a similar mammal may have. The plesiadapines were found in both Europe and North America. They and the direct ancestors of the prosimians probably had very little competition in their terrestrial low-branch niche since the rodents did not start to become really successful until the Eocene, the following epoch.

The Eocene not only witnessed the expansion of rodents into early primate niches but also the disappearance of land bridges between continents. The primates made their all-important shift to arboreal niches and were forced by strong selective pressures to adopt appropriate morphological patterns. It is very likely that the independent acquisition of arboreal traits occurred many times. It is also likely that primates were evolving very rapidly in the Eocene. There was apparently a larger primate population in the Eocene than there is today (see Fig. 9–1). Selective pressures from rodent competition and fellow primates caused multiple and rapid adaptive experimentation.

The important Eocene primates exhibited more of the traits one finds in the primate pattern—larger brains, larger and more forwardly placed eyes, and a more forwardly positioned foramen magnum. One of these fossil primate groups was the **Adapines** (*Adapidae*). Like the plesiadapines, they inhabited both Europe and North America. In Europe, these creatures were representd by a number of lemurlike primates, the best known being *Adapis*, whose body proportions and size were probably similar to those of the living lemur of today. *Adapis* lacked the prosimian dental comb. It probably did have opposable first digits. Some of these Adapines were likely ancestors of the modern lemurs and lorises. In North America we have chiefly two well-known lemuroid primates—*Notharctus* (Fig. 9–3) and *Smilodectes*. Dentally these creatures were unlike the *Adapis* but otherwise were very similar. *Smilodectes* remarkably resembled the sifaka type of indris.

The Eocene also saw a family of fossil tarsiers in Europe and North America. These tarsiers showed a considerable amount of variation but tended to have larger forebrains and shorter faces than the lemurs. *Necrolemur* (originally misnamed) was common to Europe and in some ways was more advanced than the living tarsier of today and in other ways more primitive. In the New World, the tarsier was represented by a creature known as *Tetonius*. Both *Tetonius* and *Necrolemur* were already vertical clingers and leapers. We unfortunately do not know whether these Eocene tarsiers evolved from lemuroid animals or directly from a treeshrew grade.

The Eocene was likely the time when the ancestors of the New World primates and Old World monkeys were undergoing parallel

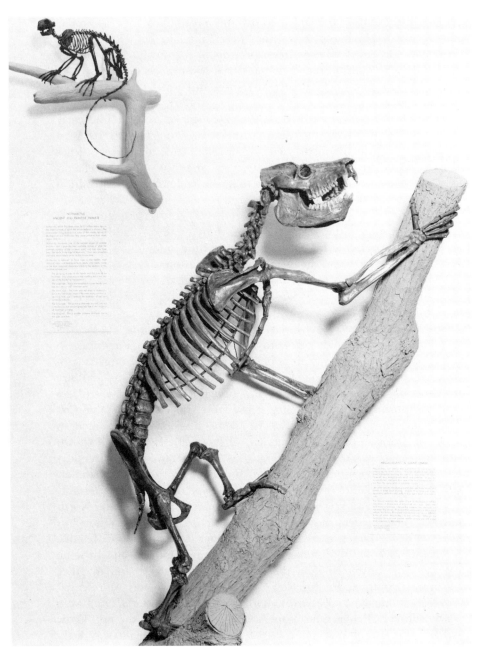

*Fig. 9–3* The small primate perched on the upper left is *Notharctus*. Note the large eyes and grasping hands and feet. The large creature is *Megaladapis,* a giant lemur, who was a calf-sized ground-dwelling form and may have lived into recent centuries, but was elminated by man. (Courtesy of the Smithsonian Institution.)

evolution. These provisional candidates for monkey status are collectively called the **Omomyids**. They inhabited North America, Europe, and part of Asia. They may even have been ancestral to the anthropoid apes. The only early candidates for ape status are (1) a creature from Burma named *Amphipithecus*, who had three premolar teeth but showed some advanced morphology, and (2) the less well-known *Pondaungia*, also from Burma.

During the Oligocene, the strong selective pressures of the Eocene resulted in fewer primate types and a smaller total world primate population. This was also the time of the first recognizable New World primates. Unfortunately, the New World primate fossils are rather incomplete. Many of them bear little resemblance to the modern New World primates. This is perhaps the most disappointing area in primate paleontology. The specimens are few and tend to be jaw fragments dating from the Oligocene, Miocene, and subfossil (recent) Pleistocene. Many exhibit some dental similarities to *Apidium*, but few resemble the modern New World primates. The oldest, *Branisella,* from the early Oligocene of Bolivia may have been related to the North American omomyids. Its teeth exhibit both prosimian and higher primate traits.

*Branisella* may be related to *Rooneyia,* an Oligocene primate from West Texas and one of only two primates known from that time in North America. *Rooneyia* showed some brain advances toward the higher primates. One South American type known as *Homunculus* may bear some relationship to the modern howler monkey, but this is not definite. Two late Miocene fossils, *Stirtonia* and *Cebupithecia,* appear to have no direct evolutionary ties to the modern New World primates. One Miocene form, *Neosaimiri,* may be the only known fossil New World primate related to a present-day creature, the squirrel monkey. Until further material is unearthed, there is no reasonable fossil structural ancestor for the modern New World primates.

The fossil-rich sands of the Fayum Depression to the southwest of Cairo, have provided a good dental ancestor of the gibbon. This Oligocene creature, *Aeolopithecus*, along with two fossil gibbons from the Miocene, provides evidence that the hylobatids have had a long and separate evolutionary history from the other anthropoid apes. It is quite possible that the hylobatids and pongids never had a common ancestor that was in any sense of the word an ape.

The two Miocene fossil gibbons, **Pliopithecus** (Fig. 9–4) and **Limnopithecus,** had not yet evolved extremely long forelimbs for brachiation. Their remains indicate they were generalized arboreal quadrupeds who probably brachiated to the extent of the modern colobines. *Pliopithecus* may have even had a tail. A creature such as *Amphipithecus* may have led to these Miocene gibbons.

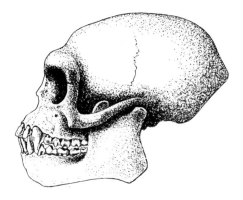

*Fig. 9–4* The skull of *Pliopithecus*. Although more generalized than the living gibbon, this creature was likely ancestral to the gibbon. (From "The Early Relatives of Man," by E. L. Simons. Copyright © 1964 by Scientific American, Inc. All rights reserved.)

An Oligocene fossil primate once thought to be a gibbon is now considered a possibility for the line leading to the hominoids. This creature is called **Propliopithecus**. Some authorities consider *Propliopithecus* too specialized for such ancestry. Unfortunately, most of the Oligocene remains are teeth and fragmentary jaws.

The clouded early fossil primate record tends to confirm that primates were very small creatures for millions of years. The populations leading to the anthropoid apes and man were probably monkeylike, at least in appearance. They were mostly arboreal creatures who occasionally experimented with terrestrial environments. In the next chapter we will look at the first recognizable great apes.

## SUMMARY

EARLY PRIMATE PATTERN
    Generalized and Small
    Insectivorelike
    Small Braincase
    Large Olfactory Development
    Eyes Lateral
    Muzzle Moderately Long
    Dentition with Rodent Features
    Slender Trunks
    Short Limbs
    Long Tails
PLESIADAPINES
*Plesiadapis*
ADAPINES

*Adapis*
*Notharctus*
*Smilodectes*
*Necrolemur*
*Tetonius*
*Omomyids*
*Amphipithecus*
*Pondaungia*
*Aeolopithecus*
*Pliopithecus*
*Limnopithecus*
*Propliopithecus*

# Chapter 10

# The Dryopithecines

## where ape and man part

Too often the lessons learned from the history of the mammals are not applied to the study of the fossil primates. Typical of the ancestry of many modern groups of mammals, the ancestors of man and the great apes might be expected to have very generalized and primitive anatomies. Such morphologically plastic primates would lead to both more specialized primates (pongids) and also to primates which would still be basically generalized (hominids). In recent years, new specimens and studies are beginning to rather accurately confirm this pattern.

The early forerunners of the great apes and possibly man are placed collectively in a group known as the **dryopithecines**. Originally only the genus *Dryopithecus* ("forest ape") was included here; this genus is known mostly from pongidlike jaws first discovered in the 1850s. These jaws possess large, projecting conical canines, diastemas, and U-shaped dental arcades. Most important is that the molar cusp pattern is characteristic of both the pongids and man. This pattern is sometimes called the **Y–5 pattern** (see Fig. 7–6). The earliest dryopithecine molars usually had five cusps, thus the "Y–5" designation. The dryopithecines have been called "dental apes."

## Aegyptopithecus

The best candidates we have for an early ancestor of the great apes is a primate from the Fayum Depression of Egypt. Found and studied by Dr. Elwyn Simons, **Aegyptopithecus** *zeuxis* is known from one skull (minus the mandible) and some jaws (Fig. 10–1). *Aegyptopithecus* dates from the late Oligocene some twenty-eight to thirty million years B.P. (before present), predating *Dryopithecus* of the Miocene. One cannot overestimate the importance of the finding of *Aegyptopithecus*. What we know of its total morphological pattern is highly significant.

*Aegyptopithecus* had the Y–5 molar pattern, and is now the oldest known fossil primate to have it. It also had large, projecting, conical canine teeth, yet the large jaws holding these teeth were part of a prosimianlike muzzle. The remainder of the skull is like that of a monkey and a lemur. A small, weakly developed sagittal crest was present (not unusual in monkeys with large dental apparatuses). The supraorbital region was moderately developed into supraorbital ridges, while the back of the skull had a well-developed occipital crest. What we

*Fig. 10–1* The skull of *Aegyptopithecus* (cast). Left—note the slight sagittal crest and occipital crest, the moderately large supraorbital ridges, the procumbent incisors, pongidlike canine, and the massiveness of the jaw. Right—from a superior view the skull has a rather lemurlike appearance. The jaw and the skull are not from the same specimen. (S. I. Rosen.)

know of the postcranial anatomy indicates that *Aegyptopithecus* was a monkey-sized primate who possessed a tail. This small-brained primate was probably a generalized quadruped. Although now a desert, the Fayum was a tropical forest area in the late Oligocene; we cannot safely label *Aegyptopithecus* as arboreal, terrestrial, or both. Whether to call this creature an ape, dental ape, or monkey is a senseless argument. What is important is that this primate is an excellent structural ancestor for the members of the genus *Dryopithecus*, ancestors of the modern pongids. *Aegyptopithecus* should probably be placed in the genus *Dryopithecus* (i.e. *Dryopithecus zeuxis*). It is likely that other Oligocene primates similar to *Aegyptopithecus* were also selectively acquiring hominidlike traits.

### *Dryopithecus*

Since the year 1856 when Edouard Lartet, a French lawyer turned paleontologist, found the first dryopithecine specimen, several hundred others have been found in Europe, Russia, India, China, the Middle East, and Africa (see Fig. 10–9). They are known to have lived throughout the Miocene and into the early Pliocene. Having existed at least ten million years and in so many places, the dryopithecines were a relatively successful group. A key to their success is found in what we know of the *Dryopithecus* morphological pattern. Although they were ancestors of the modern pongids, they were less specialized anatomically than their descendants.

*Dryopithecus*, of course, exhibited the Y–5 molar tooth pattern. This cusp pattern was less complex than the one we see in the present great apes. Large, projecting canines and diastemata were the rule, yet these large-jawed creatures usually lacked the simian shelf, a trait later acquired by the pongids. What we know of the remainder of the skull is from a creature found at East African sites. First named **Proconsul africanus**, after a famous chimpanzee at the London Zoo, this creature today is classified as **Dryopithecus africanus** and considered a likely ancestor for the chimpanzee (Fig. 10–2). The skull of Proconsul, except for the dental apparatus, is rather like a monkey's. It lacks a well-developed supraorbital region; there are no massive browridges or a torus. In fact, the frontal region is well rounded and filled out in the two Proconsul skulls that have been found. The lower face is prognathous, but the entire face does not have the dished-out appearance of the modern pongids. The occipital area possesses a moderately high and gracile (slender) crest. The known postcranial skeleton of Proconsul has

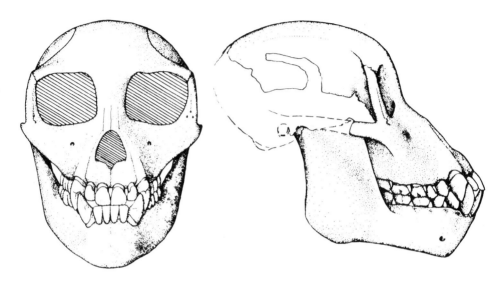

*Fig. 10–2* The skull of *Dryopithecus africanus* (Proconsul). Note the pongid jaws and dentition yet the rounded monkeylike cranial vault and supraorbital region. (After Robinson, 1952, *American Journal of Physical Anthropology*.)

indicated to most experts an upper limb that was not yet that of a brachiator. I have long felt this to be in fact an upper limb adapted for brachiation. Recently Dr. O. J. Lewis, an English anatomist, has shown that Proconsul did have the wrist of a brachiator. As to how much and how often it brachiated, we do not know. Significantly, this specialization had been evolved (at least morphologically) by the Miocene.

The East African sites where *Dryopithecus africanus* has been found should not be accepted as the limit of this chimpanzeelike creature's geographic distribution. This volcanic area is well suited for fossilization. South Africa at this time may have been forested and Proconsul may have been there also. In fact, we cannot rule out the existence of Proconsul wherever members of the genus *Dryopithecus* have been found.

A possible ancestor of the gorilla existed in the early Miocene along with Proconsul. This creature is known as **Dryopithecus major**. Its candidacy for gorilla status is largely based on its size. Both *Dryopithecus major* and Proconsul were very sexually dimorphic like the living gorilla but unlike the chimpanzee.

As to dietary niches of the African dryopithecines, we have no information. Their dentitions would have been suitable for frugivorous as well as herbivorous regimens.

From the famous Siwalik Hills fossil beds of India and Pakistan come jaws and teeth of Asian dryopithecines known as **Sivapithecus indicus**. These primates are known from the late Miocene and early Pliocene. Similar material has also been found from the early Miocene of East Africa. Likenesses to the present-day orang-utan are found in some features of the teeth and jaws; chimpanzee and gorilla features are also present. It is considered that *Sivapithecus* (more properly **Dryopithecus indicus**) is a likely ancestor of the orang-utan. *Dryopithecus indicus* may have also been present in East Africa. Its relationship to the African pongids is not clear.

The dryopithecines of Europe (*Dryopithecus fontani*) are in some traits like *Dryopithecus indicus*. These creatures were probably so generalized that one could not categorize them as chimpanzees or orang-utans. One cannot assume that the first dryopithecines evolved in Africa simply because our earliest known finds are from there. We do know that during the Miocene, Africa and the Middle East were connected by a landbridge that would have allowed for genetic interchange between continental populations. It would be significant if Charles Darwin were right about the first ancestors of man originating in Africa, but the full picture is still a long way from being completed. It does seem reasonable that the disappearance of the landbridge and the deforestation in Africa and Euroasia created narrower econiches and subsequent selective pressures for more specialized pongids—chimpanzee, gorilla, and orang-utan.

### Gigantopithecus

For hundreds of years, oriental drugstores have sold ground-up fresh and fossil teeth and bones, so-called "dragon bones." These have been marketed as potions of all sorts. In the year 1935, the Dutch paleontologist G. H. Ralph von Koenigswald found a most interesting "dragon tooth," a molar belonging to some type of fossil hominoid. Von Koenigswald believed this to be giant pongid, which he called **Gigantopithecus**. By 1939, he had found four such molar teeth but no jaws or other skeletal remains. The teeth are believed to be of middle Pleistocene age from fossil beds in Southern China. This means that *Gigantopithecus* would have been a contemporary of the true fossil hominids we will consider in the next several chapters.

The 1950s saw the first discovery of *Gigantopithecus* mandibles in China—the Kwangsi mandibles. In 1968 another lower jaw was found, this time in India and dating from the middle Pliocene. This indicates that primate populations of this type existed from perhaps five million years B.P. up to possibly 250,000 years ago. The early part of the Pleistocene is known for megafauna—unusually large or giant animals which today are much smaller. The gigantopithecines of the Pleistocene likely represent one primate version of this gigantism. There were some giant baboons in East Africa approximately two million years ago at the time of australopithecine populations. The Pliocene *Gigantopithecus* jaw is not particularly large, but it may also be a female specimen. Sexual dimorphism appears to have been present, as evidenced in the Chinese specimens.

The largest *Gigantopithecus* mandible dwarfs even that of a modern large adult male gorilla jaw (Fig. 10–3). One must be somewhat cautious in assessing total body size on the basis of teeth and jaws alone. As you will see in the next chapter, the australopithecine fossil hominids had both larger teeth and jaws than modern man, yet were in total body

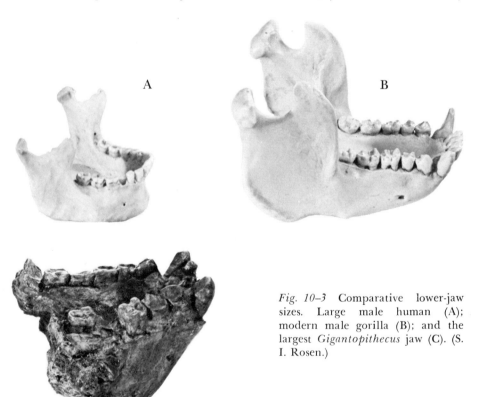

*Fig. 10–3* Comparative lower-jaw sizes. Large male human (A); modern male gorilla (B); and the largest *Gigantopithecus* jaw (C). (S. I. Rosen.)

*Fig. 10–4* An artist's conception of what *Gigantopithecus* may have looked like. (Reconstruction by R. F. Zallinger.)

size quite small. Mosaic evolution seems to have always been going on in the higher primates. It has been suggested by Dr. Simons that *Gigantopithecus* weighed some 600 pounds and was perhaps nine feet tall, dwarfing the modern male gorilla (Fig. 10–4); this idea is interesting yet so far unverifiable. These large simian-shelved mandibles probably did belong to gorillalike creatures that were evolutionary dead ends. They may have been descendents of a creature such as *Dryopithecus indicus*.

*Gigantopithecus* canines were not excessively projecting or large. Its molar teeth in some ways are rather hominid (Fig. 10–5). These features led Dr. Franz Weidenreich, one of the great students of fossil man, to deduce that this primate was in fact a giant hominid, which he

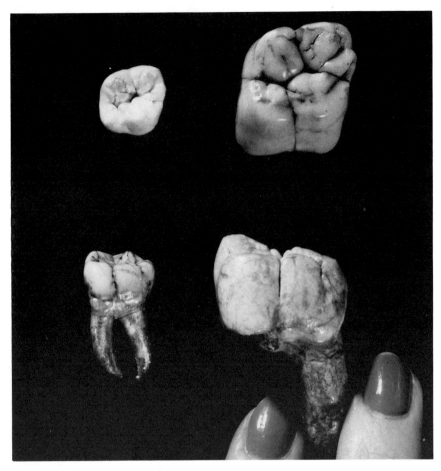

*Fig. 10–5* Left—crown and lateral views of a modern human molar tooth. Right—the same views of one of the *Gigantopthecus* molars. Note the similarity in cusp pattern between the two. (Courtesy of the American Museum of Natural History.)

called "Giganthropus." Weidenreich believed that man's early ancestors were giants. This idea has been refuted by the fossil evidence down to this day. Present popular dogma considers *Gigantopithecus* to be a giant gorillalike pongid who was ground dwelling and herbivorous. It has been suggested that the small-object feeding (T-complex) adaptation was taking place in this primate, thus the presence of some "hominid" traits.

It is evident that the early dryopithecines were very plastic (generalized) creatures that followed a number of evolutionary pathways, some of which were extinction.

*Ramapithecus*

If one adheres to the still popular idea of a pongid ancestor for man, the dryopithecines are the logical place to seek it. In 1910, G. E. Pilgrim reported on an upper-jaw fragment from lower Pliocene deposits of the Siwalik Hills. He classified this material as *Dryopithecus punjabicus*. In the early 1930s G. E. Lewis, a Yale paleontology student at the time, brought forth from the same region similar upper-jaw fragments which he classified as *Ramapithecus brevirostris*. Lewis claimed hominid status for this creature who lived at least ten million years ago. The scientific community did not accept Lewis's thesis. Beginning in 1961, Dr. Elwyn Simons began a review of the jaw fragments found by Pilgrim and Lewis; Simons also studied similar material discovered by Lewis. He concluded that all these fossils belonged to one type of primate whose appropriate taxon is **Ramapithecus punjabicus**. Simons also reissued the claim of hominid status for these creatures.

THE *Ramapithecus* PATTERN

Our knowledge of the total morphological pattern of *Ramapithecus* is confined to teeth and is very fragmentary. **Small canines** and **incisors** are evidenced in the empty tooth sockets of this creature; a small diastema may also have been present, a trait common in young hominids but rare in adults. If these front teeth were small, they were also small relative to the posterior teeth (premolars and molars). It is also claimed that these incisors and canines were vertically positioned rather than slightly procumbent as in pongids. *Ramapithecus* also has **hominidlike molars** somewhat between the dryopithecines and the australopithecines, the group we will consider in the next chapter. In light of the molar teeth of *Gigantopithecus* being hominidlike in crown pattern, we must reserve judgment until more complete specimens of *Ramapithecus* are found.

Dr. Simons has reconstructed the upper-jaw **dental arcade** of *Ramapithecus* in a smooth, **parabolic** curve, as in modern man (Fig. 10–6). These jaws are so fragmentary that they could also be reconstructed in a pongid U-shaped arcade. Some monkeys in fact do have very parabolic dental arcades, thus a total pattern is necessary even more in the case of *Ramapithecus*. It has also been claimed that this creature had a **short face**. While all of these claims may well be quite correct, we just know too little of the total morphological pattern of *Ramapithecus* to make accurate judgments.

At the same time (1961) Dr. Simons was working on the *Ramapithecus* material from India, Dr. L. S. B. Leakey discovered jaw fragments

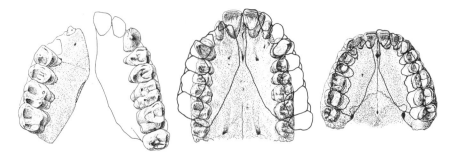

*Fig. 10–6*  Left—the upper jaw of *Ramapithecus* as reconstructed; center—compared with an orang-utan; and right—with modern man. (From "The Early Relatives of Man" by E. L. Simons. Copyright © 1964 by Scientific American, Inc. All rights reserved.)

from a primate at Fort Ternen, Kenya. Leakey quickly named this creature **Kenyapithecus** *wickeri*. Professor Simons and his coworker Dr. David Pilbeam consider this creature to be an African example of *Ramapithecus (Ramapithecus wickeri)*. This specimen dates at fourteen million years B.P.—late Miocene or early Pliocene—thus it is possibly older than the Indian finds. Candidates for *Ramapithecus* status also exist from China, Southern Germany, and Spain. *Ramapithecus* likely had a very wide geographic distribution similar to *Dryopithecus*.

It has been suggested that if *Ramapithecus* in fact did have small, nonprojecting canines and a small face, he must have also been a tool-using, bipedal primate. Such speculation is based upon the commonly held idea that long, projecting canines were selected against and lost in man's pongid ancestor in favor of erect posture, bipedal locomotion, and the subsequent use of the hands for tool-making and holding weapons for defense. Such dogma has been losing support, especially since there is no evidence for an ancestor with large canines. If *Ramapithecus* is indeed an early hominid and not a more advanced type of dryopithecine, the canine-tooth hypothesis is invalid. Obviously, we need much more complete material of this nature.

### Oreopithecus

The last early fossil primate we will look at is considered by most authorities to be an evolutionary dead end, yet this creature may be the most important fossil primate we have. The creature is **Oreopithecus** *bamboli*, first found in Tuscany, Italy, in 1872 in what was once a forested swamp but now is a coal-mining area. *Oreopithecus* has on occasion been called the "swamp ape," even though few would consider it an ape.

We have knowledge of *Oreopithecus* from some fifty individual specimens found from 1954 on by Dr. Johannes Hürzeler. This creature dates from some ten million years B.P. (late Miocene). It presents the most unusual mosaic of primate anatomical traits known. Its brain was chimpanzee sized, but its skull had a nonpongid appearance. The cranial vault is smooth and well rounded in contour, rather like some monkeys. A heavy supraorbital torus overhangs a very vertical and short face (Fig. 10–7). The nasal region has a well-developed hominid nasal bridge. The canine teeth are somewhat projecting but are not large when compared to those of apes and many monkeys. The incisor teeth are small and vertical. The entire dentition has the hominid trait of being what we call "balanced"—that is, the anterior and posterior teeth are proportionate in size to one another. Diastemata do not regularly appear, nor does the lower jaw have a chin or a simian shelf. The upper molar teeth resemble those of some fossil apes but not the living apes. The lower molars show some pseudocercopithecoid features. The crushed skull bases of *Oreopithecus* may indicate a foramen magnum and occipital condyles more forwardly placed than in the modern apes.

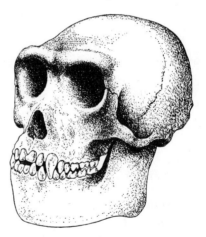

*Fig. 10–7* The skull of *Oreopithecus*. Note the well-rounded cranial vault yet large supraorbital torus. (From "The Early Relatives of Man" by E. L. Simons. Copyright © 1964 by Scientific American, Inc. All rights reserved.)

The upper limb of *Oreopithecus* is quite long, like that of a brachiating ape or monkey; yet the iliac blade portion of the pelvis is horizontally expanded in a hominid manner. This tends to indicate that *Oreopithecus* likely stood erect and moved bipedally on the ground while being a brachiator of some sort in the forest swamp. Some foot bones exhibit hominid traits, yet this primate probably had an opposable big toe.

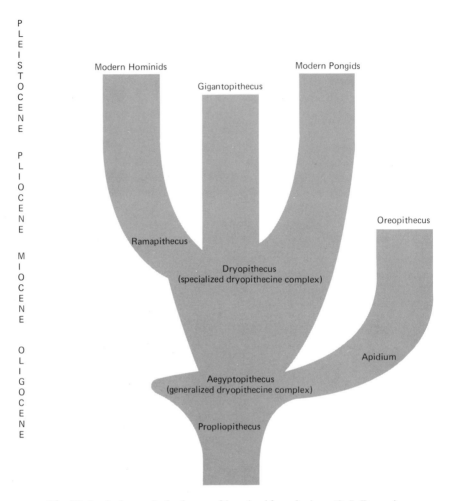

P L E I S T O C E N E

P L I O C E N E

M I O C E N E

O L I G O C E N E

Modern Hominids

Gigantopithecus

Modern Pongids

Oreopithecus

Ramapithecus

Dryopithecus
(specialized dryopithecine complex)

Apidium

Aegyptopithecus
(generalized dryopithecine complex)

Propliopithecus

*Fig 10–8*   A theoretical scheme of hominoid evolution. (S. I. Rosen.)

Many years ago, William King Gregory pointed out that the molar teeth of *Oreopithecus* are almost identical to those of an Oligocene primate known as ***Apidium***. In recent years, Dr. Simons has reaffirmed Gregory's observation. While *Oreopithecus* was likely not a direct descendent of *Apidium*, he may well have been derived from the same family.

Few authorities would classify *Oreopithecus* as cercopithecoid, pongid, or hominid. Most favor placing this primate in the superfamily *Hominoidea*. While Hürzeler and Dr. William L. Straus, Jr., the two men who have studied *Oreopithecus* extensively, tend towards placing this creature as an aberrant hominid, most students favor classifying it in its own family. Clearly *Oreopithecus* was aberrant and an evolutionary dead end.

If *Oreopithecus* is so unusual a primate, why is it so important? Because this creature establishes and makes evident that hominid, pongid, and cercopithecoid traits were probably acquired (as well as lost) independently a number of times in the experiments of primate evolution. *Oreopithecus* further drives home the principle that no animal can be defined or classified by isolated single traits; only the total morphological pattern is definitive. The lessons to be learned from *Oreopithecus* are many.

It is evident that during the Miocene and probably also the early Pliocene, higher primate populations were evolving along a number of adaptive pathways. Some would remain generalized arboreal quadrupeds while others would become brachiators. Some of these brachiators by virtue of their large size would become savannah woodland and grassland dwellers. Finally some were evolving along the hominid pathway. A few aberrant creatures such as *Oreopithecus* were adapting into multiniches; such adaptation would likely have required a more complex

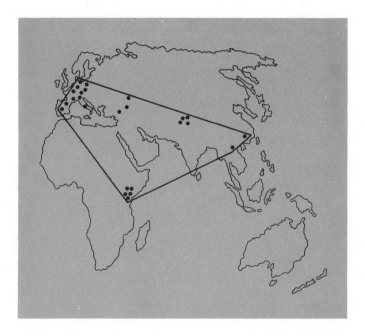

*Fig. 10–9*   Distribution of dryopithecine sites. Enclosing lines indicate theoretical distribution.

neurological organization than any of the other fossil primates possessed at this time. These potential ancestors of man would likewise have been aberrant compared to the other fossil primates. The types of primate populations that were ancestral to all these creatures would appear monkeylike but would not be definable as monkeys or apes due to their generalized anatomies. In the next chapter we will look at creatures we can say definitely had some role in human evolution.


## SUMMARY

DRYOPITHECINES
Y-5 PATTERN
AEGYPTOPITHECUS
PROCONSUL
*Dryopithecus africanus*
*Dryopithecus major*
*Sivapithecus*
*Dryopithecus indicus*
*Gigantopithecus*
*Ramapithecus punjabicus*
THE *Ramapithecus* PATTERN
      Small Canines and Incisors
      Hominidlike Molars
      Dental Arcade Parabolic
      Short Face
*Kenyapithecus*
*Oreopithecus*
*Apidium*

# Chapter 11

# The Australopithecines

The time period we will be concerned with in the remaining chapters is the Pleistocene—the time of the great ice ages, or glacial periods. The Pleistocene has seen to date four great glacial periods which in Europe are known as the Günz, Mindel, Riss, and Würm (Fig. 11–1).[1] Between these ice ages were warmer periods known as interglacials. There have been three of these, and we may now in fact be living in the fourth interglacial of the Pleistocene.

The beginning of the Pleistocene in certain parts of Europe was marked by the appearance of modern mammals such as the cow, the horse, and the elephant. These are collectively known as Villafranchian fauna; we arbitrarily term the beginning of the Pleistocene the Villafranchian, even in places where such animal life did not exist. Unfortunately the great ice ages did not occur in Africa, where many of the very early hominid fossils have been found. In Africa, the boundary between the Pliocene and the Pleistocene is not clearly marked. There has been a

[1]This is an oversimplified scheme of Pleistocene chronology.

| DIVISION | GLACIAL PERIOD | YEARS AGO |
|---|---|---|
| LATE PLEISTOCENE | Würm (late) Würm (early) | 0 100,000 |
| MIDDLE PLEISTOCENE | Third (Riss/Würm) Interglacial | 200,000 |
| | Riss | 300,000 |
| | Second (Mindel/Riss) Interglacial | 500,000 |
| | | 700,000 |
| EARLY PLEISTOCENE | Mindel | 800,000 |
| | First (Günz/Mindel) Interglacial | 1,000,000 |
| | Günz | 2,000,000 + |

*Fig. 11–1*   The Pleistocene.

practice of pushing the Pleistocene back in time as older African hominid fossils are found. The Pleistocene is now believed to have begun some two to three million years ago, and the first recognizable fossil hominids were creatures of the this epoch.

### Man-Apes and Ape-Men

No biological discoveries have stimulated twentieth-century imaginations more than the findings of these first hominids—the **australopithecines**. These creatures have at times been incorrectly called man-apes or ape-men; they are clearly early man and perhaps no closer to apes than we ourselves are. These fossil finds have been the subject of continuous controversy and remain so to this day.

The early fossil primate record, as well as almost the entire history of mammalian evolution, indicates a rather evident pattern: the earliest ancestors of a mammal group are anatomically very generalized animals. Some descendents of these biologically plastic ancestors remain generalized and others become specialized. The specialized evolutionary offspring are usually doomed to extinction because they adapt to very narrow econiches that are vulnerable to change. The biologically plastic descendants are usually able to survive, and survival often entails evolving into other forms (species).

The above pattern is rather evident in the case of the australopithe-cines, although the generalized ancestor has not positively been found. The best candidate we have for this generalized form is a fossil hominid known as *Australopithecus africanus*—the small, or *gracile australopithe-cine*. The specialized hominid is *Paranthropus robustus*, the *robust australopithecine*. Paranthropus is classified by some authorities as *Aus-tralopithecus robustus*. As mentioned before, the meaningful taxonomic level in primate paleontology is the genus as was evident in our con-sideration of the earliest fossil primates.

### *Australopithecus africanus*

**THE SITES**

The year 1924 is a most important date in the history of biology. It was in this year that Dr. Raymond Dart, an anatomist at the University of the Witwatersrand in South Africa, discovered the existence of fossil man in Africa. In 1925 Dart presented to the scientific community of the world the partial skull and natural brain cast of the first known australopithe-cine, a creature he called *Australopithecus africanus* ("southern ape, Africa"). It would have been a significant find even if it had been only a fossil ape, since no apes today live in Southern Africa, which is largely deforested. At first Dart thought this creature was intermediate between the apes and man. Upon careful examination and keen judgment, he recognized that this in fact was something more—a possible ancestor for man.

The skull was found at a limestone quarry near the **Taung** rail-way station in Botswana, Africa, in the Transvaal area. The fossil skull, known commonly as the Taung Child (Fig. 11-2), is that of a young gracile australopithecine. The first permanent molar teeth had erupted in its jaws so that on the basis of modern man's dentition, this individual would have been about six years old. We of course are not certain what the rates of growth and maturation were in fossil man. One might logically expect the earliest fossil hominids to have matured somewhat faster than modern man but more slowly than apes.

While we do not normally define a fossil type on the basis of immature specimens, fortunately for Dart and all of us, the Taung speci-men exhibits the important diagnostic traits of the gracile austra-lopithecine.

The scientific community did not readily accept Dart's pronounce-ments on *Australopithecus*. While some scholars did acknowledge hom-

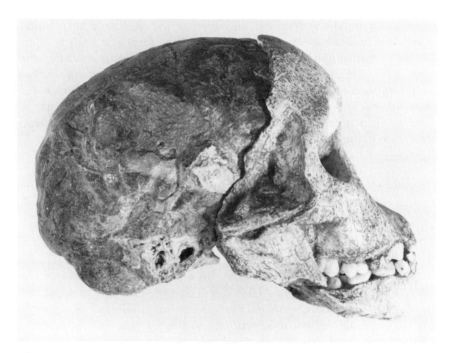

*Fig. 11–2* The actual fossil specimen from Taung, *Australopithecus africanus.* (Courtesy of P.V. Tobias; Photo by A. R. Hughes.)

inid traits in the skull, *Australopithecus* was generally viewed as some kind of aberrant ape. Superficially all australopithecine skulls, gracile and robust, appear pongid at a glance; but, as you will see, their patterns are decidedly hominid.

As Darwin had his Huxley, so Raymond Dart had his Broom—Dr. Robert Broom, a world famous Scottish paleontologist and physician. From 1936 to 1948, Broom collected in the Transvaal adult specimens of both gracile and robust australopithecines from limestone caves. Chief among these finds was an adult *Australopithecus* believed by some to be a female; this specimen was found at a site known as **Sterkfontein** (Fig. 11–3). This complete skull, except for a missing lower jaw, confirmed the anatomical pattern seen in the Taung Child (Fig. 11–4). Unfortunately, Broom had that common paleontological illness of giving each find a new taxonomic designation, in this case *Plesianthropus transvaalensis.* This creature is now classified as *Australopithecus africanus.* Additional *Australopithecus* material was found at a site called **Makapansgat**, some 150 miles north of Johannesburg. Broom also found at two other sites in the Transvaal the first known robust aus-

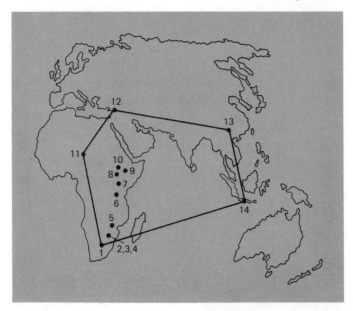

*Fig. 11–3* Some major australopithecine sites: (1) Taung, (2) Sterkfontein, (3) Swartkrans, (4) Kromdraii, (5) Makapansgat, (6) Olduvai Gorge, (7) Peninj, (8) Kanapoi, (9) Koobi Fora, (10) Omo, (11) Koro Toro, (12) Tell Ubeidiya (Israel), (13) *"Hemanthropus"* (China), and (14) *"Meganthropus"* (Java). (Modified after Brace, C. L., *The Stages of Human Evolution.* © 1967 by permission of Prentice-Hall, Inc.)

tralopithecine *(Paranthropus robustus)* specimens. Due largely to the work of Broom and his then assistant, Dr. J. T. Robinson, we have samples of both sexes and general age groups of australopithecines from South Africa.

The dating of these South African finds has always been a problem, since the modern technology of chronometric dating cannot be applied to these formations. Dating by types of animal life found at sites has presented more problems than answers. We do know that the gracile australopithecine sites in South Africa are considerably earlier than the robust australopithecine sites. Sterkfontein and Makapansgat are the oldest sites in South Africa, dating some two to three million years B.P.; this would make *Australopithecus* a creature of the late Pliocene and early Pleistocene.

Tools known as Oldowan pebble tools have been found at an extension of the Sterkfontein site. These pebble tools are the oldest man-made implements known. They are very simple tools; and yet we cannot be sure that they were in fact made by *Australopithecus*, since these tools have never been found in direct association with fossil finds

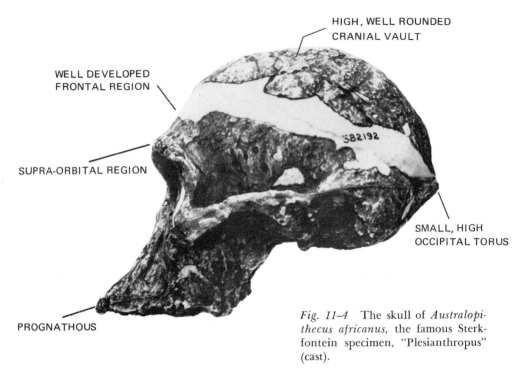

HIGH, WELL ROUNDED
CRANIAL VAULT

WELL DEVELOPED
FRONTAL REGION

SUPRA-ORBITAL REGION

SMALL, HIGH
OCCIPITAL TORUS

PROGNATHOUS

*Fig. 11–4* The skull of *Australopithecus africanus*, the famous Sterkfontein specimen, "Plesianthropus" (cast).

in South Africa; and wherever they have been found there have always been signs of other types of hominids, perhaps more advanced ones. Some scholars believe that *Australopithecus* did not have a large enough brain ("cerebral rubicon") nor an advanced enough one to fashion even these simple implements. It is felt by some that these tools were made by a more advanced hominid such as *Homo erectus*. Even with such disagreement, from what we know of nonhuman primate behavior, *Australopithecus* certainly would have been able to use these tools and probably would have been capable of making them.

### THE *Australopithecus africanus* PATTERN

The total morphological pattern of the gracile australopithecine is decidedly hominid. A number of cranial features alone establish hominid status. First, the skull of *Australopithecus* shows a surprisingly **well-developed forehead** for an early hominid. Some frontal-lobe brain expansion likely occurred in these gracile australopithecines. The **supra-orbital region** is **weakly developed**; there are no robust supraorbital ridges or a barlike torus. There is also an **elevated** and **well-developed cranial vault** (braincase); the major portion of the brain was well above the level of the face, not below and behind as in the pongids. The estimated **cranial capacity** range is **450–600 cc.**—an ape-sized brain. Yet one must remember that brain size is not an absolute indicator of neurological development. Unfortunately the natural brain casts we have of *Australopithecus* are too vague to tell us anything of real evolutionary importance. Relative to his **small body size** (he was perhaps not over four feet tall and weighed no more than eighty pounds), *Australopithecus* had a slightly larger brain than a great ape, yet absolutely it was of pongid size.

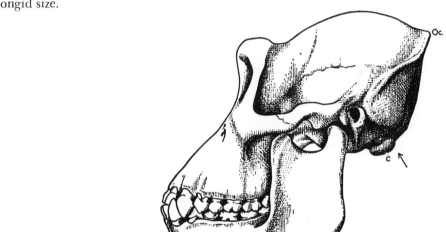

A

*Fig. 11–5* Skull of an adult female gorilla (*A*) compared with that of *Australopithecus* (*B*). Note the contrasts in the position of the occipital protuberance (*Oc.*), the inclination of the foramen magnum, the height of the braincase above the level of the supraorbital margin, the degree of development of the supraorbital region, and the relative position and inclination of the occipital condyle (*c*). (After LeGros Clark, *The Fossil Evidence of Human Evolution*, 1964, by permission of the University of Chicago Press.)

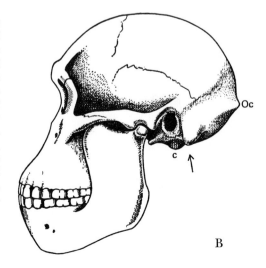

B

The nuchal muscles of *Australopithecus* were not the heavy types found in pongids. This is evidenced by a **low-placed occipital torus** on the back of the skull. This bar of bone running across the back of the skull marks the height and, to a certain degree, the nature of the neck musculature. In the case of *Australopithecus africanus* this torus is not an especially robust structure. This morphological evidence partly indicates erect posture, since heavy neck muscles are required to support a large skull that is not adequately balanced atop the spine. Further evidence of erect posture is the quite **forwardly placed occipital condyles and foramen magnum** (Fig. 11–5).

*Australopithecus* had relatively **large jaws**, which imparted a pongid appearance to the skull. In addition its **jaws** were **extremely prognathous**, even more than those of the robust australopithecine. This protrusion is probably due to the fact that the gracile australopithecine had large incisor and canine teeth that required considerable buttressing by bone. These jaws supported a quite hominid dentition. The teeth were arranged in a **parabolic dental arcade** approaching that of modern man (Fig. 11–6). While the **molar teeth** were **very large** compared to ours, so were the anterior teeth; all these teeth are very hominid in character though not identical with those of modern man. The canine tooth is very robust compared to the strangely small canines of the robust australopithecine. In size the anterior teeth of *Australopithecus* are proportionate to the posterior teeth (premolars and molars); he had a **balanced dentition.**

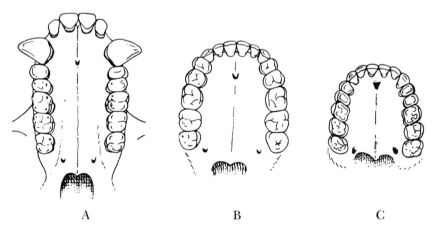

A                              B                              C

*Fig. 11–6*  The palate and upper teeth of (*A*) a male gorilla, (*B*) *Australopithecus*, and (*C*) modern man. Note that in the curved contour of the dental arcade, the small canine teeth, and the absence of a gap (diastema) between the canine and incisors, the australopithecine pattern is fundamentally of a hominid type. (After LeGros Clark, 1965. Courtesy of the Trustees of the British Museum.)

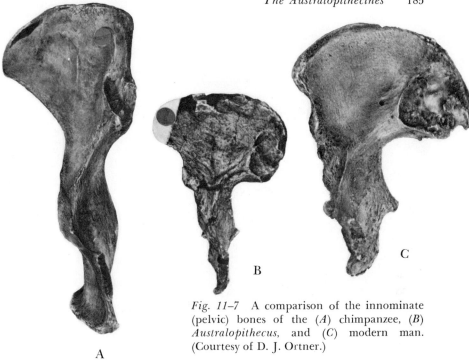

*Fig. 11–7*  A comparison of the innominate (pelvic) bones of the (*A*) chimpanzee, (*B*) *Australopithecus*, and (*C*) modern man. (Courtesy of D. J. Ortner.)

While the dental apparatus of *Australopithecus* was large, the muscles of mastication that supported it were not excessively large. **No sagittal crest** was required to anchor the temporal muscles. Another chief muscle of mastication, the masseter, was also not excessively large, thus the **zygomatic region** (cheek bone) was **moderate** in development. This latter feature is also related to the nature of the chewing forces that were absorbed in the zygomatic region.

The fairly adequate amount of postcranial material we have for *Australopithecus* dramatically affirms that the gracile australopithecine was indeed an **erect** and **bipedal** hominid. The iliac portion of the pelvis was broad and short, one of the chief characteristics of a bipedal hominid (Fig. 11–7). This does not mean that *Australopithecus* was a biped in the same sense that modern man is. His walk was probably not as perfected as ours. He may have been in fact a better runner than a walker.

The total morphological pattern of *Australopithecus* is eloquent testimony to the original conclusions of Raymond Dart and also to the soundness of our present anatomical knowledge of what an early ancestor of man would have been like. *Australopithecus africanus* is certainly a good structural ancestor for modern man, if not in fact a true one.

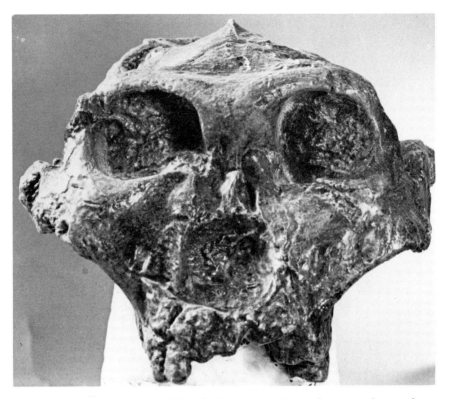

*Fig. 11–8*  The skull (cast) of *Paranthropus robustus* from Swartkrans, South Africa. (S. I. Rosen.)

## Paranthropus robustus

### THE SITES

The indefatigable Robert Broom also made the first robust australopithecine finds. These creatures, which he named **Paranthropus robustus**, lived in South Africa from the beginning of the middle Pleistocene, a million or more years later than *Australopithecus africanus*. Robust australopithecines are known from two sites in South Africa, again in the Transvaal. In 1938, *Paranthropus* was found at **Kromdraii** and in 1948 at **Swartkrans**, an older site near Sterkfontein (Fig. 11–8).

In 1959, Mary Leakey, wife of L. S. B. Leakey, found a most important australopithecine skull. Dr. Leakey, trained as an archaeologist, claimed that this was a new type of early hominid; unfortunately he named this creature **Zinjanthropus** *boisei* (Fig. 11–9). Students of fossil man today agree that Zinjanthropus is in fact an East African variety of *Paranthropus robustus.* Zinjanthropus was found in the **Olduvai Gorge** in Tanzania. The gorge is one of the great fossil beds on this planet. Today it is a vast, arid, inhospitable place; but several million years ago is was a well-watered area of one or more large lakes with lush vegetation in the surrounding areas. Early man at Olduvai may have been a lakeside dweller at times. This locale contrasts dramatically with the semiarid australopithecine sites of South Africa, which were likely savannah grasslands with some sparse woodlands.

Olduvai Gorge is geologically divided into five beds, Bed I being the oldest and deepest, Bed V the shallowest and most recent. Zinjanthropus was found in Bed I. What makes this find so important (other than proving that at least the robust type of australopithecine lived elsewhere in Africa and was variable) is the date derived for it. A radiometric dating technique (potassium-argon dating) places Zinjanthropus

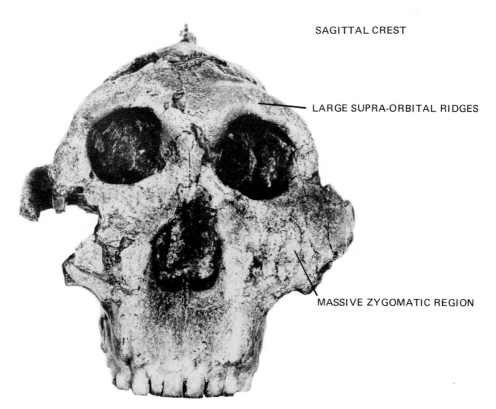

SAGITTAL CREST

LARGE SUPRA-ORBITAL RIDGES

MASSIVE ZYGOMATIC REGION

*Fig. 11–9* The skull of Zinjanthropus from Olduvai Gorge, Tanzania, East Africa. (Courtesy of D. J. Ortner.)

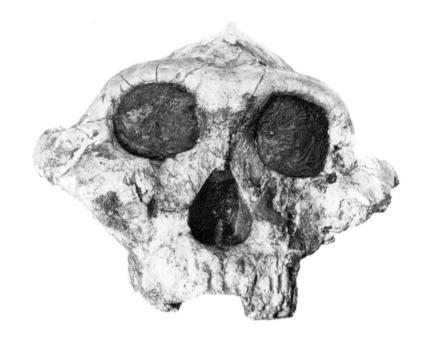

at Olduvai some 1.8 million years ago. Thus *Paranthropus robustus* was a contemporary of *Australopithecus africanus*.

In 1968 and 1969, Dr. Leakey's son, Richard, found further remains of *Paranthropus robustus* in East Africa; this time in **East Rudolf**, Kenya, at the **Koobi Fora** site. One of the specimens is a magnificently preserved skull of *Paranthropus* that is remarkably similar to Zinjanthropus (Fig. 11–10). In 1971, he also found at East Rudolf what I believe to be a female *Paranthropus*. This material from East Rudolf is possibly 2.6 million years old.

*Paranthropus robustus* material is also known from Java and China. Several early Pleistocene mandibles from Java were originally classified as **Meganthropus** *paleojavinicus*. A good number of students recognize these jaws as being probable Far Eastern forms of *Paranthropus*. There are also some drugstore teeth from China classified as **Hemanthropus** *peii* that may be *Paranthropus*. Again a wide distribution is indicated for the australopithecines.

THE *Paranthropus robustus* PATTERN

The total morphological pattern of the robust type of australopithecine can best be described as that of an **aberrant hominid**. *Paranthropus* is without doubt a hominid, but one with pongid traits; it exhibits **some gorilloid adaptations**. It must be made clear that *Paranthropus* was no more closely related to the gorilla than you or I are. It so happened that the robust australopithecine adapted to an econiche that was either similar to the gorilla's or at least required gorilla traits. *Paranthropus* was decidedly a hominid. Just the opposite may be the case with *Gigantopithecus*—an aberrant gorilla evolving some hominid traits due to T-complex adaptation. Like *Gigantopithecus*, *Paranthropus* was an **evolutionary dead end**.

*Paranthropus* was a robust creature compared to the gracile australopithecine. His height likely exceeded four feet but not by too much. His weight at its maximum was in the area of 120 pounds. *Paranthropus* also had a **large skull**, considerably larger than that of *Australopithecus* but of approximately the same cranial capacity range of **450–600 cc.** (Fig. 11–11). We thus have a creature with a lower brain-to-body weight and body size ratio than *Australopithecus*.

The frontal region of the skull of *Paranthropus* had **large supraorbital ridges**; a **forehead** was **lacking**. The cranial vault was not well rounded and expanded as in the gracile form. A moderately sized **sagittal crest** was present, at least in the male, as was a **massive zygomatic region** and **massive jaws**. The occipital torus was slightly higher up than

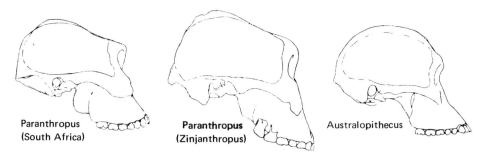

Paranthropus
(South Africa)

Paranthropus
(Zinjanthropus)

Australopithecus

*Fig. 11–11* A comparison of the skulls of *Paranthropus* from South and East Africa with *Australopithecus*. Note especially the heights of the braincase above the level of the supraorbital regions and the amount of brain space relative to skull sizes—lateral views. (From *Olduvai Gorge*, ed. L. S. B. Leakey. Vol. II: *The Cranium of Australopithecus [Zinjanthropus] boisei* by P. V. Tobias, © 1967 Cambridge University Press.)

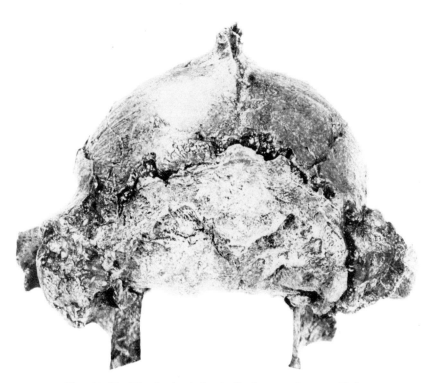

*Fig. 11–12* The back of the skull of *Paranthropus* (Zinjanthropus.) Note the sagittal crest and the extent of the nuchal area. (Courtesy of D. J. Ortner.)

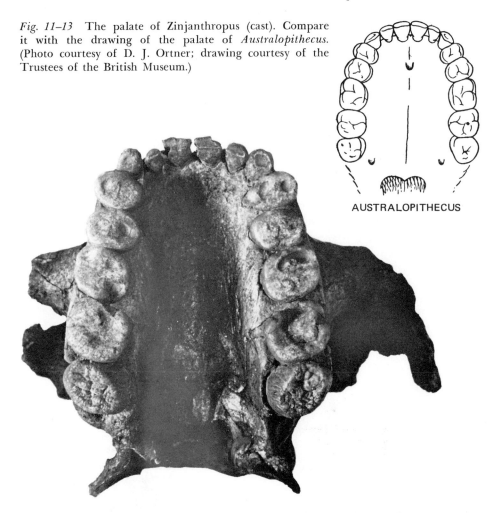

*Fig. 11–13*   The palate of Zinjanthropus (cast). Compare it with the drawing of the palate of *Australopithecus*. (Photo courtesy of D. J. Ortner; drawing courtesy of the Trustees of the British Museum.)

**AUSTRALOPITHECUS**

that of *Australopithecus*; it was more like an occipital crest. These is some indication that the neck muscles of *Paranthropus* were more extensive than in the gracile form (Fig. 11–12).

The foramen magnum and occipital condyles of the robust form were slightly more forwardly placed on the skull base than were those of *Australopithecus*. This is probably related to the **jaws** being **not very prognathous**. This latter trait is likely related to the **anterior teeth** being **quite small** when compared to the **huge molar teeth** (Fig. 11–13). Thus *Paranthropus* had an **unbalanced dentition**. The teeth were arranged in a **parabolic dental arcade** which, however, was more U-shaped than one would expect to find in a hominid. The massive cheek teeth and nonprojecting jaws of *Paranthropus* may be adaptations to small-object feeding (T-complex). *Gigantopithecus* exhibited the same features. The entire dentition of the robust australopithecine was designed for extreme crushing and grinding.

In 1964, at **Peninj**, a site west of **Lake Natron**, East Africa, a large mandible was found. This lower jaw is believed to be like that of Zinjanthropus (Fig. 11-14). Like the other mandibles of *Paranthropus,* a modified **simian shelf** is present. This structure is not placed as low as in the pongid type but is still best described as a shelf. This structure served as an additional buttress in the heavy mastication of tough materials that were probably similar to the items in the modern gorilla's diet.

What we do have of the postcranial anatomy of *Paranthropus* seems to indicate a strange mixture of gorilloid and hominid traits. The hand bones indicate a good possibility that the robust australopithecine may have been a **knuckle-walker** of some sort; the pelvis somewhat confirms this. While generally too technical a subject for this book, the pelvic

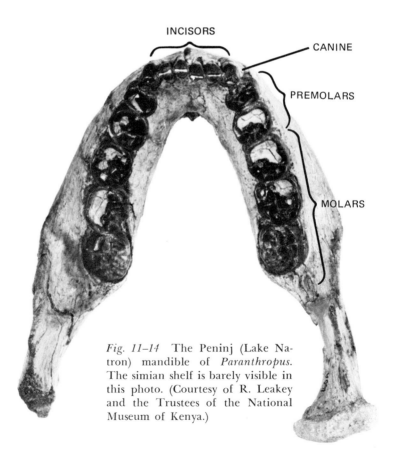

*Fig. 11-14* The Peninj (Lake Natron) mandible of *Paranthropus.* The simian shelf is barely visible in this photo. (Courtesy of R. Leakey and the Trustees of the National Museum of Kenya.)

remains we have from Swartkrans show significant differences from those found of *Australopithecus* at Sterkfontein and Makapansgat. *Paranthropus* is best described as having a **gorilloid-hominid pelvis**. In addition, features of an ankle bone from Krombraii show that *Paranthropus* possibly possessed a **divergent big toe**. The bipedalism of this creature was likely very awkward compared to that of modern man and even that of *Australopithecus*. *Paranthropus* was thus an **inefficient biped** whose locomotor pattern varied between a crude bipedalism and sporadic knuckle-walking. Yet the robust australopithecine was probably a better biped than either the chimpanzee or gorilla. It must be kept in mind that although not a pongid, a good number of the traits seen in the postcranial skeleton of this creature are related to arboreal adaptations. For example, knuckle-walking is a result of a forearm and hand whose flexor muscles are adapted for a hook-grip posture. It is not unreasonable to assume that *Paranthropus* was some type of **tree climber** as well as a terrestrial hominid. Dr. J. T. Robinson seems to generally concur with this latter opinion. Interestingly, *Paranthropus* is the only australopithecine that qualifies for the description William King Gregory gave of the australopithecines—"man-ape."

### Time-Expanding Sites

The middle and late 1960s saw several new sites and specimens which have shed new light on the chronology of the australopithecines. In 1965, the end of a humerus (upper-arm bone) was found at **Kanapoi** in Northwest Kenya, near Lake Rudolf. This site may be 3 to 4.5 million years old. This bone fragment may be an example of *Australopithecus*; but it has been debated because the distal (end) portion of the humerus in the chimpanzee and man are so similar. Unfortunately, we really know nothing about what the anthropoid apes were like during the last several million years. If this was an australopithecine, this group probably had a Pliocene history. Another site in Kenya, **Baringo**, may also have yielded australopithecine material dating three million years B.P.

One of the most exciting newer australopithecine sites is the **Omo** site in the Omo Basin of Ethiopia, to the north of Lake Rudolf. Dr. F. Clark Howell has found jaws of *Paranthropus* and some teeth and part of an *Australopithecus* skull here. This event is possibly quite significant; it proves that the two australopithecine types were probably contemporaries. The evidence is not yet conclusive, but the Omo material may be as old as four million years! Thus we have a Pliocene site. I believe that it is likely that the earliest Pliocene forms of the australopithecines were so generalized that *Paranthropus* would not have been as distinguishable from *Australopithecus* as it was in the late Pliocene and early Pleistocene. The Omo area holds great promise of solving some of the riddles of australopithecine prehistory.

At **Lothagam**, Kenya, again near Lake Rudolf, a mandibular fragment has been found that may belong to *Australopithecus africanus*. If so, this is the oldest australopithecine to date—some 5.5 million years B.P. Lothagam, as well as the other time expanding sites, confirms what many students of fossil man have felt in recent years—that is, the australopithecines had their origins in the Pliocene. These sites also demonstrate one of the frustrating aspects of primate paleontology—the discoveries raise more questions than answers.

### *"Homo habilis"*

In 1960, the Leakeys found at Olduvai Gorge in Bed I some 300 feet away and 2 feet deeper than Zinjanthropus hominid fossil remains which were first called the "Pre-Zinj Child." Later in 1964, Dr. Leakey, along with Professor Philip Tobias and Dr. John Napier, placed this material in a new hominid taxon, **Homo habilis**. These three anthropologists not only claimed that this was a new type of fossil hominid, more advanced than *Australopithecus africanus*, but also redefined the genus *Homo* to accommodate *Homo habilis*, which means "able or capable man." Dr. Leakey at this point took the Oldowan pebble tools away from Zinjanthropus and gave them to *Homo habilis*, claiming him to be the true tool-maker at Olduvai.

The *Homo habilis* material from Bed I consists mainly of two juvenile parietal (cranial vault) bones, some hand bones, a mandible, a clavicle, two leg bones, and an almost complete foot. A cranial capacity of 680 cc. has been calculated from the two cranial bones. This estimate has been challenged on the grounds that no accurate estimate of cranial capacity can be made from only these two bones.

The hand bones show some gorilloid features and may in fact belong to *Paranthropus*; the hand would have been able to achieve a hominid grasp. The collar bone is very robust and again probably belonged to *Paranthropus*. Although slightly crushed, the mandible shows great similarity to that of *Australopithecus africanus* (Fig. 11–15). The leg bones are those of a hominid—which one we are not sure. The left foot is an important find in that it represents new evidence of the postcranial anatomy of early hominids. Although perhaps flat-footed or suffering from "fallen arches," this is the foot of an erect, bipedal hominid. A grasping toe was not present here.

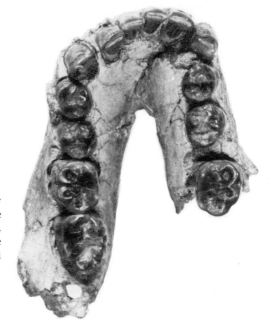

Fig. 11–15 The crushed lower jaw of "*Homo habilis.*" Note the balanced nature of the dentition. (Courtesy of R. Leakey and the Trustees of the National Museum of Kenya.)

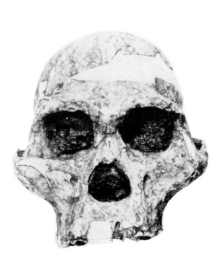

Fig. 11–16 Frontal view of the reconstructed skulls of *Australopithecus* (minus lower jaw) and *Paranthropus* (Zinjanthropus). (Courtesy of D. J. Ortner.)

Skeletal material from Bed II was also claimed to be that of *Homo habilis,* but opinion now is that this is some type of more advanced hominid *(Homo erectus).* A few very fragmented finds elsewhere in Africa have also been given possible *Homo habilis* status. Olduvai Gorge remains the main source of such material.

**Table 11–1.**    Australopithecine Total Morphological Patterns

| Trait | Australopithecus | Paranthropus |
|---|---|---|
| Cranial capacity | 450–600 cc. | 450–600 cc. |
| Cranial vault | high and expanded | low-placed and not expanded |
| Sagittal crest | absent | present in male |
| Temporal fossa | medium sized | very large |
| Occipital region | low-placed torus | low-placed but rather extensive crest |
| Frontal region | rounded and well-developed | underdeveloped |
| Supraorbital region | lack of development | large supraorbital ridges |
| Jaws | large | massive |
| Simian shelf | absent | present |
| Zygomatic region | moderately large | massive |
| Dentition | balanced | unbalanced |
| Dental arcade | parabolic | parabolic, but less so |
| Foramen magnum | in hominid but not modern position | even more forwardly placed |
| Pelvis | hominid | gorilloid and hominid |
| Foot | nondivergent big toe | divergent big toe |

## The Major Controversies

### "HOMO HABILIS"

From 1925 to this day, the australopithecines have been the center of continued debate and speculation. From the above description of *Homo habilis,* one can make a case for classifying this material into the taxon *Australopithecus africanus.* Most students of fossil man believe *Homo habilis* to be either (1) an East African version of *Australopithecus,* (2) a slightly advanced *Australopithecus,* or (3) are suspending judgment until more material (especially cranial bones) is found. It is always best to either withhold taxonomic judgment or tentatively place the find in the established taxon it most closely resembles.

ONE GENUS OR TWO GENERA

One of the major australopithecine debates has been over whether the gracile and robust australopithecines are two biologically distinct creatures or close variations of the same australopithecine theme—that is, two separate genera or two species of the same genus. Dr. J. T. Robinson, who has found and studied more of the australopithecines than any other living paleontologist, is the prime adherent of the two separate genera school. A number of years ago Robinson offered an interpretation of the australopithecine material that has become known as the **Dietary Hypothesis**. He claimed that the differences in the dentitions of the two australopithecines represented major dietary differences. The gracile form, *Australopithecus*, was an omnivore, eating meat as well as vegetation and was thus a **gatherer-hunter** in savannah grassland. The relatively large canine teeth of *Australopithecus* were a carnivorous adaptation. *Paranthropus* was a pure vegetarian, eating vegetation such as roots, bulbs, and seeds. This kind of vegetation often has silica (sand) naturally woven into its fibers and would have caused the extreme wear patterns and scarring seen on the molars (*Paranthropus* lived in both savannah woodland and grassland). Robinson based this hypothesis partially on evidence that *Australopithecus* lived in rather arid times when vegetation would have been less abundant and animal protein a more reliable food source. *Paranthropus* was thought to have lived in wetter times when vegetation would have been more lush and plentiful. We are still not completely sure what the weather was like during these times, but we do know now that these two creatures were contemporaries, at least up until two million years ago. Even if Robinson's climatic judgments prove to be wrong, the Dietary Hypothesis may hold true for adaptation to econiches, one rather general and the other more specialized. One argument favoring Robinson's position of having two separate genera is what is known as the **Competitive Exclusion Principle**—two similar creatures (species of the same genus) could not both survive for long in an environment in which they were competing for the same econiche. One would either become extinct or evolve into another niche, likely becoming a new species or genus. Similar creatures can live in the same environment and ostensibly seem to occupy the same econiche but actually be adapted to less obvious microniches that would not be apparent in the fossil record—for example, diurnal and nocturnal niches. Robinson's best argument still remains the significant morphological differences between the gracile and robust forms.

While most authorities adhere to the classification of these forms into *Australopithecus africanus* and *Australopithecus robustus*. I believe that these creatures should be maintained in the two separate genera they were originally placed in. The morphological distance between these australopithecines merits such taxonomic distinction.

## THE SEXUAL DIMORPHISM DEBATE

One of the most novel interpretations of the australopithecine record is what might be called the one-species school of thought. The leading proponent of this position (and one of its very few followers) is Professor C. Loring Brace. This thesis holds that we are not observing the remains of two major types of australopithecines but simply the male *(Paranthropus)* and the female *(Australopithecus)* of one single species. This theory is based primarily on the idea that early hominids were more sexually dimorphic than modern man, who only exhibits moderate sexual dimorphism; this is not substantiated by our fossil record to date. It is also based on the terrestrialism-sexual dimorphism idea. If *Paranthropus* was a tree climber—that is, a partially arboreally adapted primate—this thesis is not applicable.

One of the serious failings of this school of thought is the aspect of probability. How likely was it to find all males at Swartkrans and Kromdraii and only females at Sterkfontein, Makapansgat, and Taung? What are the odds of a female primate having such a different dentition from its male counterpart? The answer: infinitesimally low. Further, the evidence of both a male and a female *Paranthropus* at the Koobi Fora site eliminates the sexual dimorphism hypothesis. Part of the problem of sexual dimorphism in the australopithecines rests in the fact that *Australopithecus* was so gracile a creature that distinguishing male from female is a quite difficult task, probably requiring complete individual skeletons, which we do not have.

## TIME AND PHYLOGENY

Originally the robust australopithecines were thought to be hominids of only the middle Pleistocene and the gracile ones of the early Pleistocene. As you have seen, the evidence now confirms a long history for both forms with their overlapping in time. *Paranthropus* existed into the middle Pleistocene and, like *Gigantopithecus*, became extinct. *Australopithecus (Homo habilis)* as far as we know did not exist later than 1.8 million years ago. The key problem has been how these forms are related. A complicating factor is the presence of *Homo erectus,* a more advanced hominid who, from dental evidence, existed perhaps back

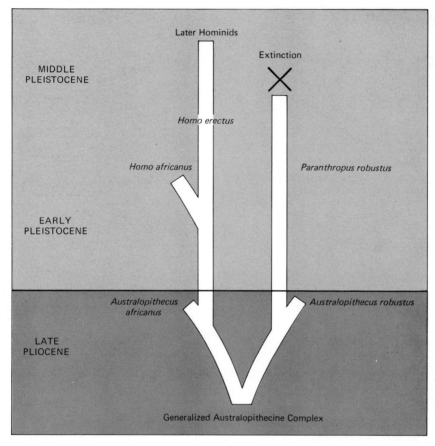

*Fig. 11–17*    The author's viewpoint of australopithecine evolution (S. I. Rosen.)

as far as 2 million years B.P. Can an ancestral form *(Australopithecus africanus)* be a contemporary of its descendant *(Homo erectus)*? One can not apply simple grandfather–grandson analogies here. These are grossly different hominids. There is the possibility that both types of australopithecines from the Pleistocene were evolutionary dead ends. In the fall of 1972, a new find from the eastern shore of Lake Rudolf, dating possibly some 2.6 million years, was announced. This incomplete skull, lacking a lower jaw and most of the face, has an estimated brain size of 800 cc. The skull was reconstructed from hundreds of fragments, so that the reconstruction is open to some question. In general conformation the skull would appear to be somewhat more advanced than *Australopithecus africanus,* especially in cranial capacity; but it is not very different from what one might expect from *Homo habilis*. Some incipient *Homo erectus* features may also be present. This find adds further weight to the idea that *Australopithecus africanus* is not in the mainline of human evolution. It also tends to confirm my ideas concerning *Homo erectus* from Java in the next chapter.

It is becoming more and more evident as additional australopithecine specimens are found that these primates probably had a wide distribution throughout the Old World. It is quite logical to expect to find some australopithecine populations that were not in the mainstream of hominid evolution. Africa may have been the "Garden of Eden," but only for a very brief time.

In light of the total morphological pattern of *Australopithecus* from the Pleistocene and its full-fledged hominid status, there is no reason it should not be placed in the genus *Homo*—that is **Homo africanus**. I am not the first to suggest this and hopefully not the last. The placement of a fossil in the genus *Homo* does not necessarily imply that that creature was a direct ancestor of ours, it only implies a general degree of similarity to us. As more australopithecines are discovered, especially from the Pliocene, their biological pattern will become apparent (Fig. 11–17).

## SUMMARY

AUSTRALOPITHECINES
*Australopithecus africanus*
GRACILE AUSTRALOPITHECINE
*Paranthropus robustus*
ROBUST AUSTRALOPITHECINE
TAUNG
STERKFONTEIN
MAKAPANSGAT
THE *Australopithecus africanus* PATTERN
> Well-Developed Forehead
> Supraorbital Region Weakly Developed
> Elevated and Well-Developed Cranial Vault
> 450–600 cc. Cranial Capacity
> Small Body Size
> Low-Placed Occipital Torus
> Forwardly Placed Occipital Condyles and Foramen Magnum
> Large and Extremely Prognathous Jaws
> Parabolic Dental Arcade
> Very Large Molar Teeth
> Balanced Dentition

Chapter 12

# Homo Erectus

Probably the most influential naturalist in the late nineteenth century was the German comparative anatomist Ernst Haeckel. He was the chief European evolutionist, and his influence is felt even today. Haeckel coined two commonly used words—"ecology" and "missing link." He was also a theoretician of human evolution. The missing link between man and the apes was a creature he called *"Pithecanthropus alalus,"* or speechless ape-man. Haeckel, along with several other scientists of the day, influenced a young Dutch anatomist who eventually set out to find this so-called missing link. This man was Eugene Dubois. He left a promising academic medical career to find a fossil man that no one, including himself, knew for sure existed. He first searched in Sumatra but left with no hominid fossils. He finally found his "ape-man" in 1891 at **Trinil**, Java. He named the creature **Pithecanthropus erectus**, or "erect ape-man." What he found at Trinil near the Solo River was a hominid calotte (skullcap) bearing unusually archaic traits, a femur (thigh bone) that was like modern man's, a jaw fragment, and some teeth (Fig. 12–1). Chemical tests later proved the skullcap and femur to be of the same age. One could only conclude that an early erect biped had been found.

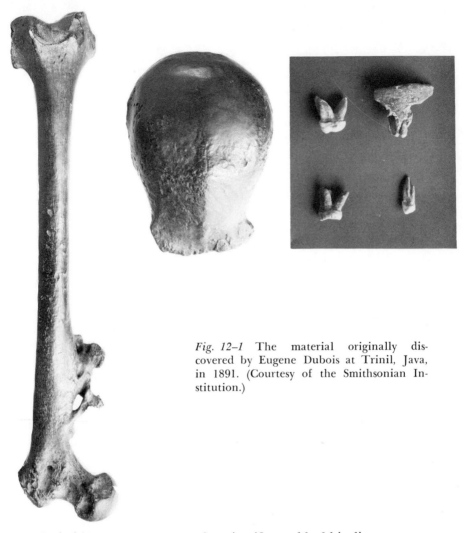

*Fig. 12–1* The material originally discovered by Eugene Dubois at Trinil, Java, in 1891. (Courtesy of the Smithsonian Institution.)

Dubois's announcement to the scientific world of his discovery met with even harsher criticism than Dart's of *Australopithecus*. The great Prussian pathologist Rudolph Virchow called the skullcap that of a giant gibbon; Virchow, an amateur anthropologist, was later found to have a perfect record in assessing fossil man—he was always wrong. Others called the remains those of a pathological being. The major question was how an "apelike" skull could belong with a human thigh bone. Today we recognize mosaic evolution, whereas even an evolutionary history for man was not widely accepted in the late nineteenth century. Eugene Dubois, one of the truly enigmatic figures in the history of science, after years of hard-fought battles over Java Man and years of retirement from the fossil battlefield, recanted his original opinions and called his find a giant gibbon.

In 1950, Dr. Ernst Mayr, the Harvard taxonomist, pointed out that it was foolish to continue to use the taxon *Pithecanthropus* when clearly these creatures were well within the genus *Homo*. We now use the taxon **Homo erectus**. Even the use of the colloquial term "pithecanthropines" is inappropriate.

## The Sites

### JAVA

Dubois's work was carried on in Java by the Dutch Paleontologist, G. H. Ralph von Koenigswald, who in the 1930s found further specimens, including some from the **Djetis** or oldest beds. The Trinil beds are of middle Pleistocene age, some 700,000 years B.P. The oldest Djetis material may be as old as 2,000,000 years. In Java we see a micro-evolutionary series for *Homo erectus,* the earlier (Djetis) material having a more archaic morphology than the later finds. Java Man was evolving over a period of some one and a half million years. It is a possibility that *Paranthropus* in the form of "Meganthropus" was contemporary at Trinil with middle Pleistocene Java Man. It is also possible that the earliest Djetis. material (teeth) may be from an australopithecine of a type similar to *Homo africanus*, which eventually became the earliest *Homo erectus* in Java. Today we technically refer to **Java Man** as **Homo erectus erectus.**

### CHINA

The first *Homo erectus* material from China was in the form of a few teeth found by a young Canadian anatomist, Davidson Black, who taught at Peking Medical College. Black studied these teeth and pronounced them to have belonged to a new type of fossil man, **Sinanthropus pekinensis.** Today we classify *Sinanthropus* as **Homo erectus pekinensis—Pekin Man.** We now have the remains of over forty individuals from the limestone cave site of **Choukoutien** (forty miles from Peking), where more extensive material was found. Unfortunately this material was lost or destroyed just prior to World War II.

After Davidson Black's untimely death, his work fell into the hands of an individual who would soon become one of the great human paleontologists of all time, Dr. Franz Weidenreich. An anatomist and physiologist by training, Wiedenreich was a well-known scholar before

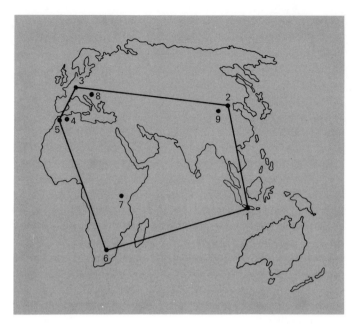

*Fig. 12–2*    The major *Homo erectus* sites: (1) Java, (2) Choukoutien, (3) Heidel-berg, (4) Ternifine, (5) Rabat, (6) "Telanthropus" (Swartkrans), (7) Olduvai Gorge, (8) Vértesszöllös, and (9) Lantian. (After Brace, C. L., *The Stages of Human Evolution.* © 1967 by permission of Prentice-Hall, Inc.)

he turned to the study of fossil man. It is Weidenreich's excellent casts of the Pekin Man material that we have today. His morphological studies of the Choutkoutien material eloquently demonstrated that Pekin Man was a slightly more advanced geographic variant of *Homo erectus* from Java.

Tools of the chopper-chopping tool tradition were found at Chou-koutien with Pekin Man. These tool types are an advanced version of the simple Oldowan pebble-tool tradition. Similar tools have been found in Java but not in direct association with the *Homo erectus erectus* finds. The Peking material is from the late Mindel glaciation—perhaps 300,000 years B.P. The lower cave at Choukoutien was a kill site—not only of animals but likely also of Pekin Man. Besides cranial fractures, the long bones of Pekin Man and other animals had been split length-wise for their marrow. Like his counterpart from Java, Pekin Man was a big-game hunter. The ashes and hearths at the Choukoutien site evidence a rather continuous use of fire. At one time this site was con-sidered the first evidence of man's use of fire. There is another site in Hungary exhibiting remains of fire that may be even older than Choukoutien.

In 1963, a new example of *Homo erectus* in China was found in Shenshi province. The evidence here is a calotte and some facial fragments. This creature is known as **Lantian Man**. It is an important find not only because it establishes *Homo erectus* in China some 600,000 years B.P. but also establishes a creature morphologically similar to *Homo erectus erectus* (Djetis) in China. A microevolutionary series in China is thus also indicated.

EUROPE

In 1907, a complete mandible was discovered at the Mauer sand pit near Heidelberg, Germany. The find was not associated with any tools. The jaw is from the first interglacial period (Günz-Mindel), some 400,000 years B.P. The **Heidelberg** or **Mauer Mandible** is huge, yet the teeth are small (Fig. 12–3). Although opinion is not uniform, this mandible probably represents *Homo erectus* in western Europe—*Homo erectus heidelbergensis*.

A more recent European find is from **Vértesszöllös**, a site near Budapest, Hungary. The date here is approximate to that of Heidelberg Man. The specimen is an occipital bone whose anatomy is that of *Homo erectus*, yet it may indicate a greater cranial capacity (1,400 cc.) than known to date for *Homo erectus*. The tools found at Vértesszöllös seem

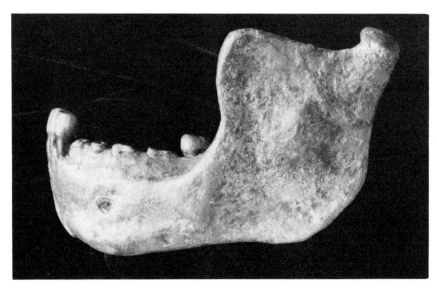

*Fig. 12–3* The Heidelberg mandible. (Courtesy of the American Museum of Natural History.)

to be a variant of the chopper-chopping tool tradition. Evidence of possible use of fire also exists at this site, predating Choukoutien. There is also dental and tool evidence of *Homo erectus* from Czechoslovakia, near Prague. This material dates from first interglacial and is called the Prezletice site.

### AFRICA

In North Africa beginning in the year 1954, mandibles, teeth, and very fragmented cranial vault material were found at two sites: **Ternifine** in Algeria; **Sidi Abderrahman** near Casablanca, Morocco; similar material had been found earlier (1933) near **Rabat**, Morocco. These remains date from middle Pleistocene times. The mandibles and teeth are considered to greatly resemble those of Pekin Man. Tools similar to those found with Pekin Man were also found. The North African finds were originally placed in the invalid taxon *"Atlanthropus mauritanicus,"* now *Homo erectus mauritanicus.*

In 1960, near the top of Bed II at Olduvai Gorge, an excellent *Homo erectus* skullcap was found. This find is commonly called **Chellean Man**, named after the types of tools found at the site. This skullcap (technically Chellean III) shows some remarkable affinities to the skulltops at Choukoutien (Fig. 12–4). This specimen is dated at 490,000 B.P.—roughly equivalent to late *Homo erectus* in Java. Near the bottom of Bed II another skullcap was discovered dating about one million years B.P. This specimen, called Maiko Gully "George," is in my opinion that of a form of *Homo erectus*, again not unlike Pekin Man.

More recently from Bed IV at Olduvai, a femur and parts of a pelvis have been found and assigned to *Homo erectus*. The femur compares favorably with the material from China at Choukoutien. Since we have no definite pelvic remains of *Homo erectus*, this half-pelvis is a quite important find if it in fact belongs to *Homo erectus* (Fig. 12–5). It belonged to an erect bipedal hominid. It does exhibit some features not found in any other known hominids. It is my opinion that the locomotor pattern here was not exactly like ours today but very close.

At Swartkrans, South Africa, Drs. Broom and Robinson not only found *Paranthropus*, but also mandibular fragments and part of the upper jaw of a creature they felt to be quite distinct from either type of australopithecine. They tentatively named this creature *Telanthropus capensis*. After reevaluating this material, Robinson expressed the opinion that this creature was in fact not a new type of fossil hominid, but represented *Homo erectus* in South Africa, a contemporary of *Paranthropus*. I generally agree with this judgment, but it should be noted

*Fig. 12–4* The Chellean Man skullcap. (S. I. Rosen.)

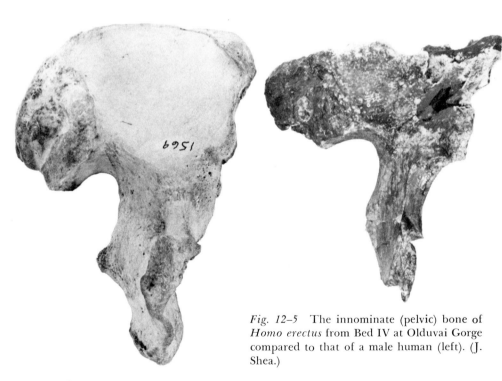

*Fig. 12–5* The innominate (pelvic) bone of *Homo erectus* from Bed IV at Olduvai Gorge compared to that of a male human (left). (J. Shea.)

that most authorities have not accepted this stand or have been waiting for additional material.

In 1969, the additional material was found not in the ground, but in a museum drawer in South Africa. What was rediscovered was the partial cranium (Swartkrans 847) that may belong to the upper-jaw fragment of *Telanthropus*. The face and browridge region are clearly not those of an australopithecine. The taxonomic affinities of this specimen may be with *"Homo habilis"* from Beds I and II at Olduvai. Whether one calls this Swartkrans hominid *Homo erectus capenensis* or *Homo habilis capenensis* is not the issue. The important matter is that we have evidence of a transitional form between some kind of early australopithecine type and the universally accepted *Homo erectus* types.

It is evident from the above sites that, again, a fossil hominid had an extensive Old World distribution. *Homo erectus* may also have been in Israel. Assuming this evolutionary grade evolved from some type of australopithecine, a similar geographic distribution would be expected.

### The Homo erectus Pattern

If one were to create a hypothetical fossil hominid that would represent a logical morphospecies (a species defined by anatomy alone) intermediate between *Homo africanus* and the hominids of the late Pleistocene, one could do no better than the pattern we find in *Homo erectus*. It is understandable why Dubois's finds were not readily accepted by the scientific community of that time. The cranium of *Homo erectus* is indeed somewhat "apeish." As in the case of the australopithecines, the pongid appearance is very superficial and only slightly obscures an even more advanced hominid.

Compared to *Homo africanus,* the skull of *Homo erectus* is broader and longer, and is set lower. The cranial capacity range is very large— **775–1,300 cc.** (Fig. 12–6). Java Man has a range of 775 to 975 cc., while the later, more anatomically advanced Pekin Man had a 850 to 1,300 cc. range. While the two *Homo erectus* subspecies overlap in brain size, it is evident that the Pekin type had taken a giant step towards modern man (1,450 cc.). Chellean Man has been estimated at 1,000 cc., which was likely the average *Homo erectus* cranial capacity over one and a half million years. How well-developed the brain of *Homo erectus* was has been the subject of some debate. Some students of fossil man have claimed the scanty impressions left on the inner cranial wall show some chimpanzee features; the plain fact is that we have never been able to obtain any real information from such impressions. What is significant is that in a period of approximately one million years the hominid brain size at least doubled from 600 to 1,300 cc. This is a fantastic increase in such a short period of evolutionary time. Some believe that the aquisition of cultural tool-making brought about a selection for a larger, more complicated brain. We naturally cannot be sure what caused this oversized brain pattern.

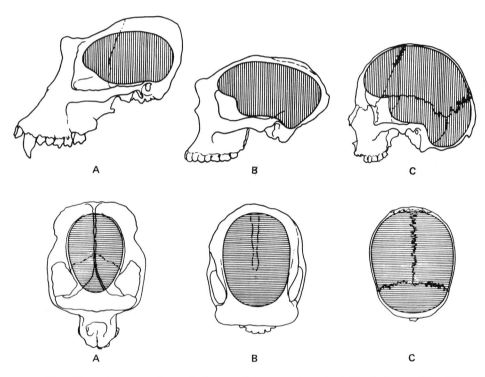

*Fig. 12–6* A comparison of the cranial capacities (shaded) of the gorilla (*A*), *Homo erectus* (*B*), and modern man (*C*). Note the degree to which the upper jaw projects beyond the supraorbital region, and the position of the brain relative to the face. (After Weidenreich, 1941. Courtesy of the American Philosophical Society.)

The top of the skull of *Homo erectus* exhibits **extreme platycephaly**, "flat-headedness" (Fig. 12–7). We find this to be less extreme in Pekin Man due to his more expanded cranial vault. The top of the cranial vault also possessed a distinct **sagittal keel**, a gable-type structure. This keel is not to be confused with a sagittal crest. The temporal muscles did not extend onto it. This trait may represent the way the outer bony table of the cranium was formed in infancy and childhood by the pull of the temporal muscles.

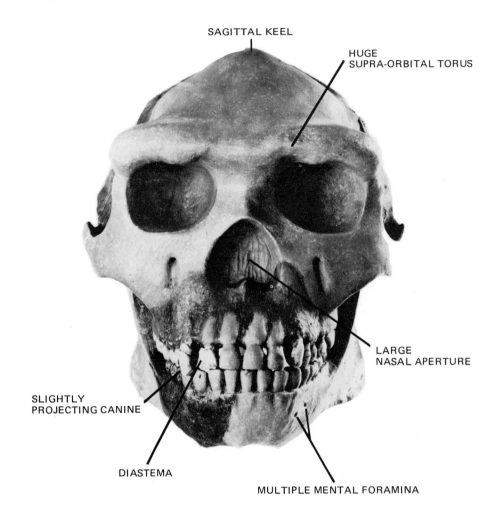

SAGITTAL KEEL

HUGE
SUPRA-ORBITAL TORUS

LARGE
NASAL APERTURE

SLIGHTLY
PROJECTING CANINE

DIASTEMA

MULTIPLE MENTAL FORAMINA

*Fig. 12–7* The skull of *Homo erectus* as reconstructed by Franz Weidenreich. (Courtesy of the American Museum of Natural History.)

The frontal region possesses a **huge supraorbital torus**—a continuous bar of bone. Pekin Man exhibits a trend towards the separation of the torus into two large supraorbital ridges. While the frontal region in Java and Pekin Man is meager, Pekin Man does show the beginning of an expansion into a true vertical forehead.

The back of the *Homo erectus* skull possesses a **massive occipital torus**, representing the upper limits and major attachment of the nuchal musculature (Fig. 12–8 and Fig. 12–9). Such musculature was requisite

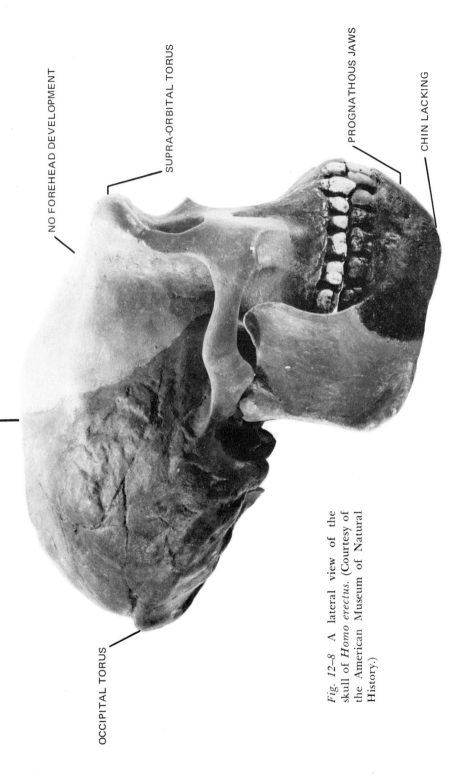

PLATYCEPHALY

NO FOREHEAD DEVELOPMENT

SUPRA-ORBITAL TORUS

PROGNATHOUS JAWS

CHIN LACKING

OCCIPITAL TORUS

*Fig. 12–8* A lateral view of the skull of *Homo erectus.* (Courtesy of the American Museum of Natural History.)

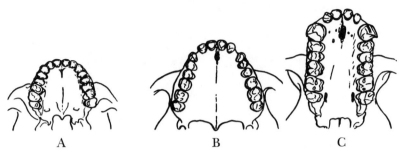

*Fig. 12–9* The upper jaws and dental arcades of modern man (*A*), *Homo erectus* (*B*), and the gorilla (*C*). Note the diastema in the jaw of Java Man. (After Weidenreich, 1941. Courtesy of the American Philosophical Society.)

for holding this very heavy head with its **large prognathous jaws** erect upon the spine. *Homo erectus* had **unusually thick cranial walls**. The **foramen magnum** was in a **modern**, centrally located **position**, indicating along with the femur that this was indeed a **habitually erect** and **bipedal** hominid, considerably advanced over the australopithecines.

The face of *Homo erectus* was broad and large. The **nasal aperture** (nose hole) was **broad**, and the nasal bridge was probably depressed as in pongids. The zygomatic bones were large relative to the total facial dimensions. In some of the Java specimens the upper jaw did have an **occasional diastema** in front of a slightly projecting and pointed canine tooth. This is not the case for the lower jaw. This diastema is not found in Pekin Man, but some Chinese jaws did have sharp, but not projecting canines. This canine trait may reflect the carnivorous nature of *Homo erectus*, since these creatures were big-game hunters. The **molar teeth** are more **advanced** in morphology over those of *Homo africanus*; however, they do exhibit some archaic and pongid traits such as wrinkling (crenulation) of the tooth enamel and **taurodontism**—large pulp cavities and fused roots in the molar teeth. The **palate** was **huge** and parabolic, but not as curved as one would expect in some specimens. **No chin** was present on the lower jaw and **no simian shelf** is evidenced. In some of the Java specimens there were **multiple mental foramina** in the front of the mandible; these are small holes which transmit vessels and nerves to the "chin" region and lower lip. They are single in each side of the jaw of modern man while multiple in pongids.

The thigh bones found at Trinil and Choukoutien evidence larger and taller creatures than the australopithecines. Java Man is estimated at five feet eight inches, while Pekin Man was around five foot one. The average *Homo erectus* stature was probably about five and a half feet. Again, there was a major evolutionary shift—an increase in stature of one and a half feet over the australopithecines and also an increase in body size and weight.

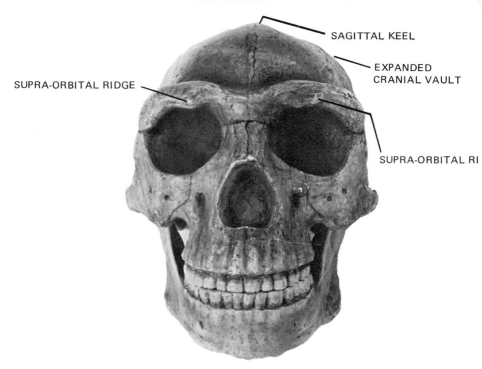

SAGITTAL KEEL

EXPANDED
CRANIAL VAULT

SUPRA-ORBITAL RIDGE

SUPRA-ORBITAL RI

*Fig. 12–10* The skull (cast) of Pekin Man. (J. Shea.)

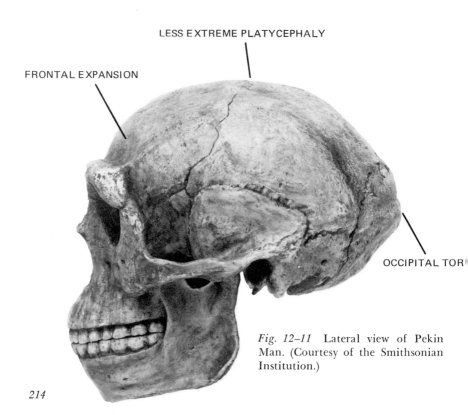

LESS EXTREME PLATYCEPHALY

FRONTAL EXPANSION

OCCIPITAL TOR

*Fig. 12–11* Lateral view of Pekin Man. (Courtesy of the Smithsonian Institution.)

**Table 12–1.** Comparative Cranial Patterns of Early and Late *Homo erectus*

| Trait | *Homo erectus erectus* | *Homo erectus pekinensis* |
|---|---|---|
| Cranial capacity range | 775–975 cc. | 850–1,300 cc. |
| Average cranial capacity | 875 cc. | 1,075 cc. |
| Platycephaly | extreme | less pronounced |
| Sagittal keel | extreme | reduced |
| Supraorbital region | huge torus | reduced, beginning to divide into separate ridges |
| Occipital torus | massive and well-defined | less massive; occipital region more expanded |
| Mandible | large | reduced, more modern |
| Chin | absent | slight indication of beginning |
| Diastema | occasional | absent |
| Canine tooth | some projecting | nonprojecting |

### The Problem of Solo Man

From 1931 to 1933 in the Solo River Valley at a site called **Ngandong** in Central Java, eleven skullcaps were found, along with two shin bones. The bases show evidence that the skulls were broken into for their contents. **Solo Man**, the creature these bones belonged to, has been called a "tropical neanderthaler" by some. What we have of his total morphological pattern seems in part more like that of *Homo erectus* (Fig. 12–12). Large supraorbital tori and occipital tori are found with a high, rounded cranial vault between them. The brain space is around 1,100 cc. Besides the morphological interpretation of Solo Man, the major problem is one of time. Solo Man is relatively late—late Pleistocene somewhere between 100,000 and 200,000 years B.P. The question naturally is why do we have Solo Man at a time when morphologically more advanced hominids are appearing over much of the rest of the Old World? It has been suggested that Solo Man represents a population that migrated via landbridges to Australia and evolved into the Australian aboriginal. He might also be an evolutionary lag, isolated from the mainstream of late Pleistocene hominid evolution. Weidenreich, who studied Solo Man and wrote a monograph that was never completed due to his death, considered the Solo population to be a form of *Homo erectus (Homo erectus soloensis)*. Some even place him as a subspecies of *Homo sapiens*.

My opinion is that Solo Man's morphological affinities rest in the *Homo erectus* direction. We have never had any other evidence of what

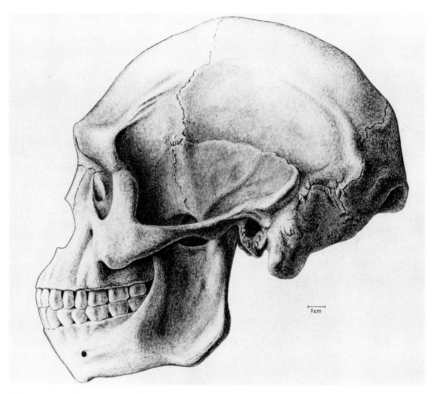

*Fig. 12–12*   The skull of Solo Man as reconstructed by Franz Weidenreich. The skullcap was the only original remains. (Courtesy of the American Museum of Natural History.)

late *Homo erectus* in Java was like. It is not unplausible that *Homo erectus,* as generally found in Java, represents a more archaic population that was isolated a good part of the time from the mainstream of *Homo erectus* evolution in Europe and mainland Asia. While total isolation from other Asian populations was unlikely, genetic interchange was probably limited at various times, especially from the beginning of the late Pleistocene. We cannot eliminate the idea that *Homo erectus* of Java was an evolutionary dead end.

The case of Solo Man illustrates an important point. Although we must use them in order to adequately communicate, so-called stages or morphological grades of organization are abstractions. We find small bits of the spectrum of past human evolution and call these "types." When we run into specimens that do not fit these concepts, we falter. If one were able to line up single-file all of the ancestors of man, we could not say where man begins and the nonhuman primate ends. The line would probably be so long that we would not be able to walk its length in a lifetime.

Except for the late Dr. L. S. B. Leakey, students of fossil man have agreed that *Homo erectus* was our direct ancestor. *Homo erectus* probably evolved from early *Homo africanus* populations in very late Pliocene times. *Homo erectus* was likely better adapted physically and culturally to the colder climates of the middle Pleistocene. He may have even hunted some australopithecines. The aberrant *Paranthropus* would have not been in competition with *Homo erectus* for the same econiche. Hominid evolution was moving very rapidly at this time; it is not unlikely that some *Homo erectus* populations would become evolutionary dead ends, and also some would linger on in isolated areas.

## SUMMARY

TRINIL
*Pithecanthropus erectus*
*Homo erectus*
DJETIS
*Homo erectus erectus*
*Sinanthropus pekinensis*
*Homo erectus pekinensis*
JAVA MAN
PEKIN MAN
CHOUKOUTIEN
LANTIAN MAN
HEIDELBERG OR MAUER MANDIBLE
VÉRTESSÖLLÖS
TERNIFINE
SIDI ABDERRAHMAN
RABAT
CHELLEAN MAN
*Telanthropus*

THE *Homo erectus* PATTERN
775–1,300 cc.
Extreme Platycephaly
Sagittal Keel
Huge Supraorbital Torus
Massive Occipital Torus
Large, Prognathous Jaws
Unusually Thick Cranial Walls
Modern Positioned Foramen Magnum
Habitually Erect and Bipedal
Broad Nasal Aperture
Occasional Diastema
Advanced Molar Teeth
Taurodontism
Huge Palate
No Chin
No Simian Shelf
Multiple Mental Foramina
NGANDONG
SOLO MAN

# Chapter 13

# The Neanderthals

It seems logical to expect that the last several hundred thousand years of the fossil record would present a simple, clear-cut picture of the evolution of modern man. This is not the case. The last part of the story is full of inconsistencies, contradictions, and puzzles. Even though we have a great many fossil specimens for this period, again, the more fossils, the more problems. The key factor in the puzzle has been what are sometimes called the **classic, or extreme neanderthals**. Morphologically these creatures are best described as the **specialized neanderthals**. These hominids first appear in the fossil record some 100,000 years ago, the latter part of the third interglacial period, and abruptly disappear around 35,000 years B.P. at the end of the first part of the last great ice age (Würm I). During the early part of the Würm, perhaps 50,000 years B.P., we also find the first anatomically modern hominids, which are collectively called **Cro-Magnon Man**. These hominids were anatomically distinct from the specialized neanderthals and yet were contemporaries of them.

The morphological and time gaps between Pekin Man and these late Pleistocene fossil hominids cover some 150,000 years and gross anatomical distances. The gaps are filled by very fragmentary finds. These specimens have collectively been called **early, or progressive neanderthals** as well as presapiens and premousterians. They are best described as **generalized neanderthals.**

## The Generalized Neanderthals

SWANSCOMBE

From 1935 to 1936 and in 1955 in a gravel pit near Kent, England, were found the parietal and occipital bones of a hominid from the late second interglacial period, about 200,000 years B.P. This skull is known as **Swanscombe**. The back of this skull is very modern in appearance, except for its great breadth (Fig. 13–1). The cranial capacity has been estimated at 1,325 cc. well within modern limits. Hand-axes and flake tools of an advanced type like the implements used by *Homo erectus* were found with Swanscombe Man. This type of generalized neanderthal was probably in an evolutionary line leading to modern man.

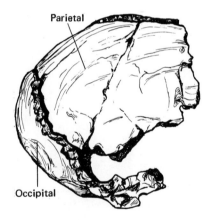

*Fig. 13–1* The fragmentary Swanscombe skull. (After Brace, C. L., *The Stages of Human Evolution*. © 1967 by permission of Prentice-Hall, Inc.)

STEINHEIM

In 1933 near Stuttgart, Germany, an almost complete but crushed skull was discovered in a gravel pit dating from about the same time as Swanscombe. This is the **Steinheim** skull. The total morphological pattern here seems to foreshadow the specialized neanderthals (Fig. 13–2).

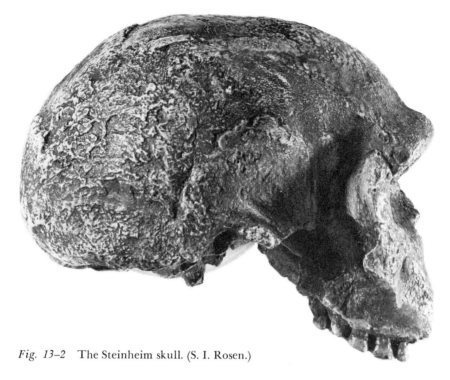

*Fig. 13–2* The Steinheim skull. (S. I. Rosen.)

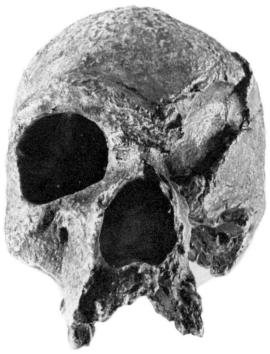

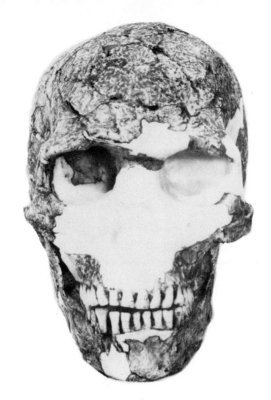

*Fig. 13–3*   The skull of Skhūl V. (Courtesy of D.J. Ortner.)

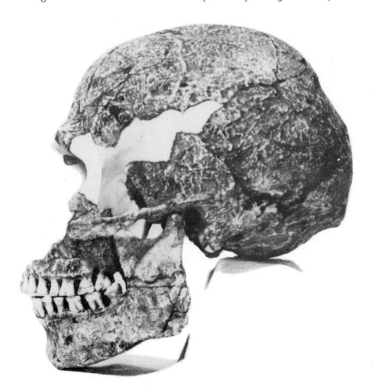

## FONTÉCHEVADE

In southern France in 1947, in a third interglacial deposit, skullcap fragments were found at a site known as **Fontéchevade**. Although not extreme, this material seems to be in the direction of modern man. Like all the generalized neanderthal finds, Fontéchevade has been the subject of much debate.

## MOUNT CARMEL

There are other generalized neanderthal finds such as Ehringsdorf in Germany (third interglacial), which appears modern, while Saccopastore, also of the last interglacial, appears to be on the way toward the specialized neanderthal in Italy. Perhaps the most unusual finds were those beginning in 1929 in Israel at **Mount Carmel**. Two fossil hominid populations living in caves were found there. Both were once thought to be of the same date, but it is now believed that the **Tabūn** population may be as much as 10,000 years (50,000 B.P.) older than the **Skhūl** population. The Tabūn remains are quite similar to the specialized neanderthals. The younger Skhūl people are for all purposes modern man (Fig. 13–3). The finding of such different morphological types so close in time and at the same location has been a preplexing problem. Some scholars have felt that the Mount Carmel site represents a place where modern man *(Homo sapiens sapiens)* was interbreeding with specialized neanderthals. Others have maintained that there were two distinct populations, indicating separate evolutionary pathways for modern man and the specialized neanderthal, since it is unlikely the Tabūn population could have evolved into modern man in so short a time as 10,000 years. Still other investigators claim that the two populations simply reflect the range of variation of late Pleistocene man.

## TAUTAVEL

More recently, a cave in the French Pyrenees has opened up new and perhaps important evidence of what the generalized neanderthals were like. Found in the village of Tautavel, these skeletal remains are known as **Tautavel Man**, who lived about 200,000 B.P. An almost complete skull was found exhibiting an interesting pattern. Of small cranial capacity, the skull appears much like *Homo erectus*. This may represent an ancestral population of the generalized neanderthals. It was in this same part of Europe that the first specialized neanderthals were discovered.

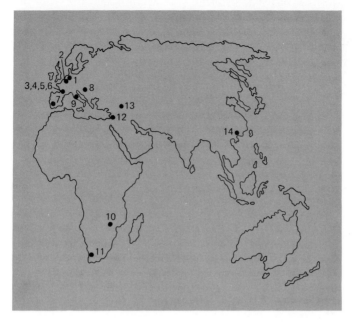

*Fig. 13–4* Distribution of the Western European (specialized) neanderthals and related non-European finds. (1) Neander, (2) Spy, (3) La Chapelle-aux-Saints, (4) Le Moustier, (5) La Ferrassie, (6) La Quina, (7) Gibraltar, (8) Krapina, (9) Saccopastore, (10) Broken Hill (Rhodesian Man), (11) Saldanha, (12) Mount Carmel, (13) Shanidar, and (14) Mapa. (After Brace, C. L., *The Stages of Human Evolution.* © 1967 by permission of Prentice-Hall, Inc.)

### The Specialized Neanderthals

We are blessed (or more correctly, haunted) by a mass of skeletal material representing the specialized neanderthals. Such material may have been discovered as long ago as the year 1700 A.D., but it did not come to the attention of scientists until the latter half of the nineteenth century. The specialized neanderthal sites are almost too numerous to list. Some of the most famous are **La Chapelle-aux-Saints; Le Moustier; La Quina** in Southwest France; the famous **Neander** Valley find near Dusseldorf, Germany; **Spy** in Belgium; and **Gibraltar** at the famous rock (Fig. 13–4). All of these finds are in Western Europe and show a basic morphological pattern.

### The Specialized Neanderthal Pattern

The Western European specialized neanderthal exhibits a total morphological pattern quite unique among hominids. These traits are not generally found among the variations of modern human populations.

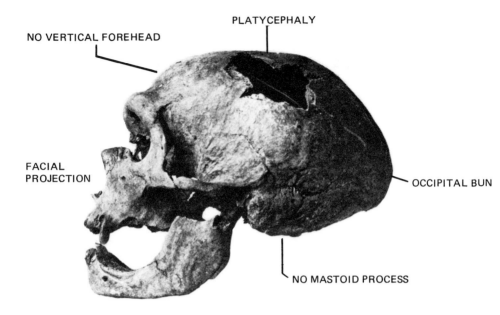

NO VERTICAL FOREHEAD

PLATYCEPHALY

FACIAL PROJECTION

OCCIPITAL BUN

NO MASTOID PROCESS

*Fig. 13–5* The actual La Chapelle-aux-Saints skull from different views. Note the premature loss of teeth and loss of bone from the jaw. (After Boule. Courtesy of the American Museum of Natural History.)

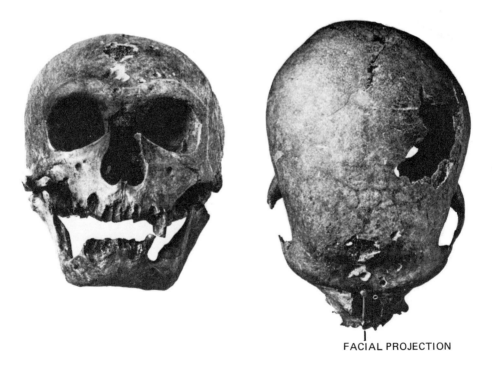

FACIAL PROJECTION

LeGros Clark believed this so strongly that he favored classifying the specialized neanderthals as a separate species of man—*Homo neanderthalensis,* the original taxon given to these hominids. Most scholars believe the specialized neanderthals would have been capable of interbreeding with Cro-Magnon Man and favor the taxon *Homo sapiens neanderthalensis.*

The specialized neanderthals had skulls which exhibit very extreme traits. **Massive supraorbital ridges** adorn the front of **extremely platycephalic skulls** (Fig. 13-5). **No forehead** is present. While the skull is of a low set, the cranial capacity range of **1,300–1,700 cc.** exceeds even that of modern man. We know little of the intelligence of these creatures, but they did have ritual burial and cultures probably as complicated as Cro-Magnon Man's, whose cranial capacity also exceeded ours. The back of the skull is grossly expanded into an **occipital bun** (or chignon) for the attachment of neck muscles. While these were erect bideds, the **foramen magnum** is **slightly deflected** from its modern

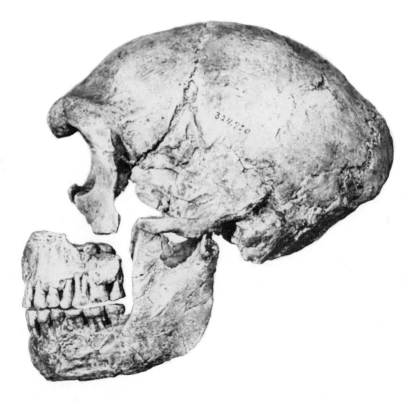

*Fig. 13–6*    The La Quina skull. (Courtesy of D. J. Ortner.)

orientation; this did not likely alter its posture or gait. **Very large projecting faces** unlike those of any other hominids are the rule (Fig. 13–6). The **jaws** were **very large**, and a **chin** was **lacking**. A very large palate held molar teeth that displayed considerable **Taurodontism**.

While the skull of the specialized neanderthal could represent an evolutionary extension of the *Homo erectus erectus* pattern, the **postcranial skeleton** is **aberrant**. It shows unusual features in the spine, shoulder blade, and pelvis that are beyond the scope of this book. Notable were the **short** but **robust** and **bowed limb bones; short in stature**, these populations were not over five feet tall as a rule. The specialized neanderthal was once thought to have had a very stooped posture due to the description of the arthritic skeleton of La Chapelle. It is now held that he was virtually as erect a hominid as we are today. I personally believe that the aberrant nature of the postcranial skeleton and some of the skull features of the specialized neanderthal would not have made for a posture identical to that of modern man. I believe that, as with *Paranthropus*, this is an **aberrant hominid**.

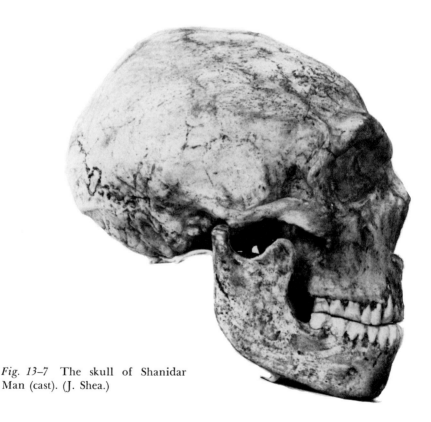

*Fig. 13–7* The skull of Shanidar Man (cast). (J. Shea.)

### Other Specialized Neanderthals

It was once thought that the extreme type of neanderthal was geographically limited to Western Europe where it was entrapped by a great ice barrier. The last several decades have witnessed new neanderthal finds outside Western Europe that exhibit affinities to the specialized type. In **Shanidar** Cave in Northern Iraq, hominids were found which for all purposes are specialized neanderthal man (**Fig. 13–7**). They also show similarities to the Tabūn material. Shanidar Man has been dated at 45,000 B.P. In 1959, the **Petralona** skull was found in Greece; it appears to be a variation of the specialized neanderthal around 50,000 B.P. From South China have come skeletal remains from the **Mapa** site, again a variation of the extreme type of neanderthal. Thus, the specialized neanderthal, or variations of this type, existed throughout the Old World in the early Würm.

### The Problem of Rhodesian Man

In 1921, at a South African site, **Broken Hill**, in Zambia, remains of two fossil skeletons were found. We call this creature **Rhodesian, or Broken Hill Man**. The pattern is strange—a specialized neanderthal skull with some Solo Man affinities and a perfectly modern postcranial anatomy (**Fig. 13–8**). In 1953, similar material was found at **Saldanha Bay**, South Africa. What is even more confusing is that this material dates about 35,000 years B.P., quite recent in time. Perhaps this is an evolutionary holdover from a *Homo erectus* population that was evolving towards the specialized neanderthal pattern late in time. Again, more questions than answers.

### The Late Omo Material

In 1967, the Omo region of Ethiopia yielded material representing three individuals dating approximately 100,000 B.P. Two of the skulls, while not complete, show quite different patterns. The **Omo I** skull appears to be quite modern, similar to the Skhūl material (**Fig. 13–9**). **Omo II** is for all purposes a *Homo erectus* type of skull rather similar to Solo Man (**Fig. 13–10**).

Every year more fossil material is being brought to view from less familiar fossil areas in Eurasia. Some of these finds show both modern

Fig. 13–8 The skull of Rhodesian Man (Broken Hill).

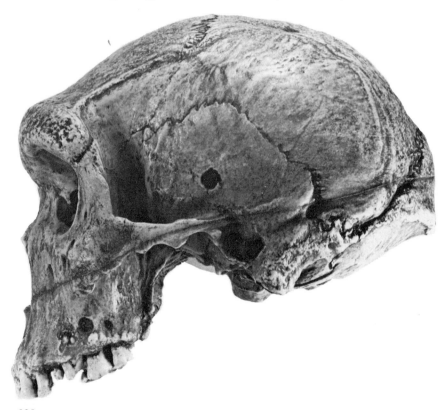

*Fig. 13–9* (Above) The Omo I skull. Note its very modern appearance. (Courtesy of M. H. Day.)

*Fig. 13–10* (Below) The Omo II skull. Note the similarities to Solo Man. (Courtesy of M. H. Day.)

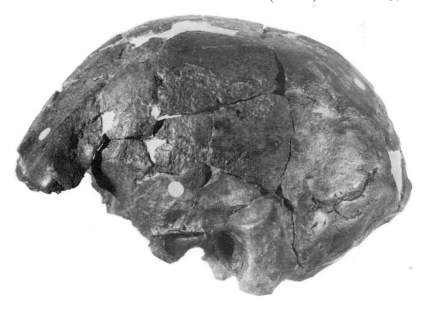

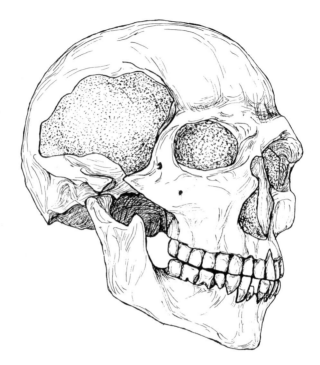

*Fig. 13–11*   One of the Předmost skulls from Czechoslovakia, perhaps 30,000 years old. Note the mixture of modern and specialized neanderthal traits. (After Brace, C. L., *The Stages of Human Evolution.* © 1967 by permission of Prentice-Hall, Inc.)

and specialized neanderthal traits (Fig. 13–11). We are finding that the tool tradition of the extreme neanderthal, the mousterian tradition, is not solely confined to Western Europe. For years it was thought since Cro-Magnon Man appears in the same places as the specialized neanderthal, he must have eliminated these creatures. This is unlikely unless he introduced some communicable diseases that the neanderthals had not been exposed to previously. It is just as likely that the specialized neanderthal interbred with Cro-Magnon Man—that is, he was genetically amalgamated into this more modern population after perhaps 70,000 years of relative isolation and adaptation to cold environments. Perhaps some 100,000 years ago hominids who would retain some specialized neanderthal traits evolved in other parts of the Old World interbreeding at times with *Homo sapiens sapiens.* Some of these populations would remain conservative in pattern (Rhodesian Man)—that is, closer to the *Homo erectus* pattern. Some *Homo erectus* populations (Omo II and Solo Man) would remain the same and thus be evolutionary dead ends.

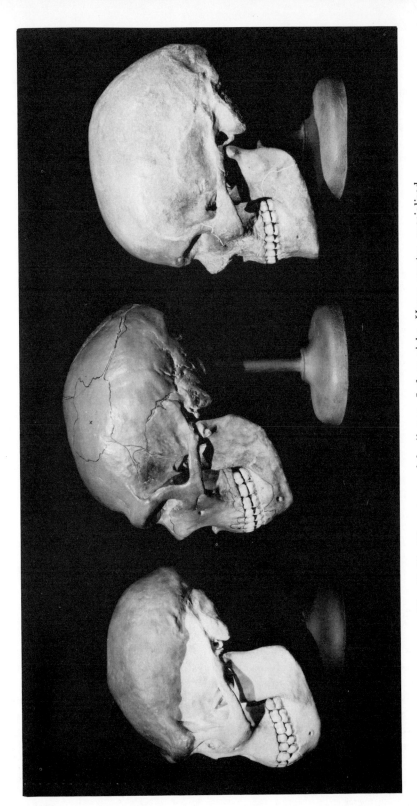

*Fig. 13–12* Three types of fossil man. Left to right—*Homo erectus*, specialized neanderthal, and Cro-Magnon Man. (Courtesy of the American Museum of Natural History.)

The logical question is why do we have only *Homo sapiens sapiens* today? We really do not know. It is unlikely that what appear to have been aberrant populations went on to become the modern races of man; man today is just not that variable. Clearly it is illogical to think of a unilinear evolution for man. While it is perhaps not a particulary comforting thought, we probably all carry some *Homo erectus*

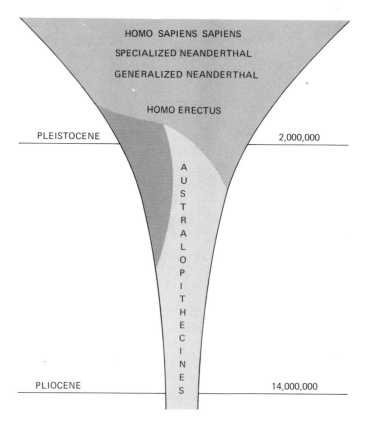

Fig. 13–13   A reasonable scheme of human evolution. (Modified after E. R. Kerley.)

and specialized neanderthal genes. If Cro-Magnon Man was the truly successful hominid we believe him to be, he was reproductively successful in terms of population numbers, and genetically absorbed the less successful. Today we are the end product of a confusing biological history in which a precise type of hominid was selected. More fossils may not give us more answers. If we knew more about the brains and behavior of the past, we would probably have our answers.

# SUMMARY

CLASSIC, OR EXTREME NEANDERTHALS

SPECIALIZED NEANDERTHALS

CRO-MAGNON MAN

EARLY, OR PROGRESSIVE NEANDERTHALS

GENERALIZED NEANDERTHALS

SWANSCOMBE

STEINHEIM

FONTECHEVADE

MOUNT CARMEL

TABŪN

SKHŪL

TAUTAVEL MAN

LA CHAPELLE-AUX-SAINTS

LE MOUSTIER

LA QUINA

NEANDER

SPY

GIBRALTER

*Homo neanderthalensis*

*Homo sapiens neanderthalensis*

THE SPECIALIZED NEANDERTHAL PATTERN

    Massive Supraorbital Ridges

    Extremely Platycephalic

    No Forehead

    1,300–1,700 cc.

    Occipital Bun

    Slightly Deflected Foramen Magnum

    Very Large Projecting Faces

    Very Large Jaws

    Chin Lacking

    Very Large Palate

    Taurodontism

    Aberrant Postcranial Skeleton

    Short and Robust Bowed Limb Bones

    Short Stature

    Aberrant Hominid

SHANIDAR

PETRALONA

MAPA

BROKEN HILL

RHODESIAN MAN

OMO I

OMO II

Chapter 14

# The Man of Tomorrow

Some of the questions frequently asked of the anthropologist by the layman are "What will the man of the future look like? Will his brain increase and his body wither away?" We have seen from the fossil record that man once had a larger brain than we have today, and seems to have become larger and taller in the last several hundred thousand years. While these questions are honest ones, they are not important, nor are the answers. The important question is whether man will be here at all.

The last five thousand years of human history have been filled with human misery. The several million years of humanity that preceded these five thousand were probably different. For most of man's pre-history, populations were small. To survive the elements of nature and time, early man would have had to cooperate to the utmost with his fellow beings. Perhaps this is the key lesson any student can learn from man's prehistory and even from the nonhuman primates.

Man of today is bequeathing the man of tomorrow a frightful legacy,—technology instead of human cooperation, polyethylene econiches, and wars designed by a few for the many. Senseless and careless use of nuclear energy may carve a pitiful niche for the man of tomorrow. Our legacy may be to inherit a nuclear wind.

234

Hopefully, in spite of his artificial specializations, man is still biologically and culturally a plastic creature; he can purposefully change. Man has learned how to biologically survive to a degree, but not how to solve the problems his culture creates. The answers do not rest with a few men, but in the cooperation of all humanity. We must solve these problems today, for we are the man of tomorrow.

# Suggested Readings

THE LIVING PRIMATES

ALTMANN, S. A. *Social Communication Among Primates*. Chicago: University of Chicago Press, 1967.

ANKEL, F. *Einführung in die Primatenkunde*. Stuttgart: Gustav Fisher Verlag, 1970.

BUETTNER-JANUSCH, J. *Origins of Man: Physical Anthropology*. New York: John Wiley and Sons, Inc., 1967.

CARPENTER, C. R. *Naturalistic Behavior of Nonhuman Primates*. University Park: Pennsylvania State University Press, 1964.

CHANCE, M. and C. JOLLY. *Social Groups of Monkeys, Apes, and Men*. London: Jonathan Cape, 1970.

DEVORE, I., ed. *Primate Behavior: Field Studies of Monkeys and Apes*. New York: Holt, Rinehart & Winston, 1965.

DOLHINOW, P., ed. *Primate Patterns*. New York Holt, Rinehart & Winston, 1972.

EIMERAL, S. AND I. DEVORE. *The Primates*. New York: Life Nature Library, Time, Inc., 1965.

JAY, P., ed. *Primates: Studies in Adaptation and Variability*. New York: Holt, Rinehart & Winston, 1968.

JOLLY, A. *The Evolution of Primate Behavior.* New York: The Macmillan Company, 1972.

KUMMER, H. *Primate Societies: Group Techniques of Ecological Adaptation.* Chicago: Aldine-Atherton, Inc., 1971.

LeGros CLARK, W. E. *History of the Primates.* Chicago: University of Chicago Press, 1965.

_____. *The Antecedents of Man.* Edinburgh: Edinburgh University Press, 1959.

NAPIER, J. R. *The Roots of Mankind.* Washington, D.C.: Smithsonian Institution, 1970.

_____. and P. H. NAPIER. *A Handbook of Living Primates.* London: Academic Press, 1967.

POIRIER, F. E. ed. *Primate Socialization.* New York: Random House, 1972.

SCHULTZ, A. H. *The Life of Primates.* London: Weidenfeld and Nicholson, 1969.

THE FOSSIL PRIMATES

BIRDSELL, J. B. *Human Evolution: An Introduction to the New Physical Anthropology.* Chicago: Rand McNally and Co., 1972.

BRACE, C. L. *The Stages of Human Evolution.* Englewood Cliffs, N.J.: Prentice-Hall, Inc., 1967.

CAMPBELL, B. G. *Human Evolution: An Introduction to Man's Adaptations.* Chicago: Aldine-Atherton, Inc., 1966.

COON, C. S. *The Origin of Races.* New York: Alfred A. Knopf, 1962.

DAY, M. H. *All-Color Guide—Fossil Man.* New York: Grosset & Dunlap, 1970.

LeGros CLARK, W. E. *The Fossil Evidence for Human Evolution,* Chicago: University of Chicago Press, 1964.

_____. *Man-Apes or Ape-Men?—The Story of Discoveries in Africa.* New York: Holt, Rinehart & Winston, 1967.

HOWELL, F. C. *Early Man.* New York: Life Nature Library, Time, Inc., 1967.

_____. and F. BOURLIERE, eds. *African Ecology and Human Evolution.* Chicago: Aldine-Atherton, Inc., 1963.

PILBEAM, D. *The Evolution of Man.* London: Thames and Hudson, 1970.

_____. *The Ascent of Man: An Introduction to Human Evolution.* New York: The Macmillan Company, 1972.

PHEIEFFER, J. E. *The Emergence of Man.* New York: Harper & Row, 1969.

SIMONS, E. L. *Primate Evolution: An Introduction to Man's Place in Nature.* New York: The Macmillan Company, 1972.

TIME-LIFE EDITORS. *The Emergence of Man.* New York: Time-Life Books, Time, Inc., 1972.

# Index

# Medway High School

599
R

Rosen, S.I.

Introduction to the Primates

599

R          Rosen, S.I.

AUTHOR
Introduction to the Primates

TITLE

                                7972

| DATE DUE | BORROWER'S NAME | ROOM NUMBER |
|---|---|---|
| MAR 19 '74 | Mary Callins | |
| NOV 12 '75 | Robert Lewis | 205 |
| MAR 17 '76 | Bob Giousnella | 201 |
| MAY 27 '77 | Steve Garner | |
| OCT 24 '78 | | |

DATE

MAR 1

NOV 1

MAR